# Aquaculture and Fish Farming

# Aquaculture and Fish Farming

Edited by
**Brendan Marshall**

Larsen & Keller
www.larsen-keller.com

Aquaculture and Fish Farming
Edited by Brendan Marshall
ISBN: 978-1-63549-031-2 (Hardback)

**Larsen & Keller**

Published by Larsen and Keller Education,
5 Penn Plaza,
19th Floor,
New York, NY 10001, USA

**Cataloging-in-Publication Data**

Aquaculture and fish farming / edited by Brendan Marshall.
      p. cm.
Includes bibliographical references and index.
ISBN 978-1-63549-031-2
1. Aquaculture. 2. Fish culture. 3. Fisheries.
I. Marshall, Brendan.
SH135 .A73 2017
639.8--dc23

The publisher's policy is to use permanent paper from mills that operate a sustainable forestry policy. Furthermore, the publisher ensures that the text paper and cover boards used have met acceptable environmental accreditation standards.

Printed and bound in the United States of America.

For more information regarding Larsen and Keller Education and its products, please visit the publisher's website www.larsen-keller.com

# Table of Contents

# Preface

This book is a compilation of chapters that discuss the most vital concepts in the field of aquaculture and fish farming. It provides in depth knowledge about the field. Aquaculture is the practice of farming molluscs, algae, crustaceans, fish, and many other aquatic animals. It has proved to be a viable alternative to the more common fishing practices that are present at the moment. Other types of activities included in aquaculture are algaculture, shrimp farming, mariculture, oyster farming, etc. The topics included in it are of utmost significance and bound to provide incredible insights to readers. The textbook aims to serve as a resource guide for students and experts alike and contribute to the growth of the discipline.

A foreword of all chapters of the book is provided below:

**Chapter 1** - Aquaculture can be defined as a sustainable practice of fish harvesting that strives to keep aquatic biodiversity and ecosystems intact. Aquaculture provides an alternative to wild fish sanctuaries throughout the world. The chapter on aquaculture offers an insightful focus, keeping in mind the complex subject matter; **Chapter 2** - Aquaculture has proven to be a very popular food system due to its commercial viability, potential for entrepreneurial development and job creating capability. It can also be adapted to any region that can support fisheries. The major components of aquaculture such as mariculture and oyster farming are discussed in this chapter; **Chapter 3** - Aquaculture has been particularly effective in the cultivation and harvest of certain marine species. Some of the more popular species that have been brought into the fold of aquaculture cultivation are seaweed, catfish, algae, giant kelp, sea cucumbers etc. The topics discussed in the chapter are of great importance to broaden the existing knowledge on aquaculture; **Chapter 4** - Since no species lives on its own, it is necessary to understand its natural habitat, its prey and its predators. Aquaculture requires the natural interaction of aquatic elements with the species that is being cultivated. The study of aquatic ecosystems provides knowledge that proves invaluable for the growth and sustenance of the farmed species; **Chapter 5** - With the growth and establishment of aquaculture as a viable source for marine food, there have evolved different systems of aquaculture practices. Some of these are aquaponics, organic aquaculture, and recirculating aquaculture systems. The major components of aquaculture systems are discussed in this chapter; **Chapter 6** - Fisheries can be defined as any region that cultivates and farms fish and other marine animals for personal or commercial use. Fisheries provide livelihood for millions of the population, especially in the third world. This chapter is an overview of the subject matter incorporating all the major aspects of fisheries; **Chapter 7** - Concerns over climate change and the propagation of biodiversity have led to the division of fisheries into sustainable fisheries and wild fisheries. The chapter strategically encompasses and incorporates the major components and key concepts of fisheries, providing a complete understanding; **Chapter 8** - Aquaculture management includes the growth and well-being of the marine species that are being cultivated. The knowledge of diseases that afflict fish and other aquatic animals are also very necessary. This chapter touches upon the various diseases and parasites that are found among aquatic animals.

At the end, I would like to thank all the people associated with this book devoting their precious time and providing their valuable contributions to this book. I would also like to express my gratitude to my fellow colleagues who encouraged me throughout the process.

**Editor**

# Introduction to Aquaculture

Aquaculture can be defined as a sustainable practice of fish harvesting that strives to keep aquatic biodiversity and ecosystems intact. Aquaculture provides an alternative to wild fish sanctuaries throughout the world. The chapter on aquaculture offers an insightful focus, keeping in mind the complex subject matter.

Aquaculture, also known as aquafarming, is the farming of aquatic organisms such as fish, crustaceans, molluscs and aquatic plants. Aquaculture involves cultivating freshwater and saltwater populations under controlled conditions, and can be contrasted with commercial fishing, which is the harvesting of wild fish. Broadly speaking, the relation of aquaculture to finfish and shellfish fisheries is analogous to the relation of agriculture to hunting and gathering. Mariculture refers to aquaculture practiced in marine environments and in underwater habitats.

Aquaculture installations in southern Chile

According to the FAO, aquaculture "is understood to mean the farming of aquatic organisms including fish, molluscs, crustaceans and aquatic plants. Farming implies some form of intervention in the rearing process to enhance production, such as regular stocking, feeding, protection from predators, etc. Farming also implies individual or corporate ownership of the stock being cultivated." The reported output from global aquaculture operations would supply one half of the fish and shellfish that is directly consumed by humans; however, there are issues about the reliability of the reported figures. Further, in current aquaculture practice, products from several pounds of wild fish are used to produce one pound of a piscivorous fish like salmon.

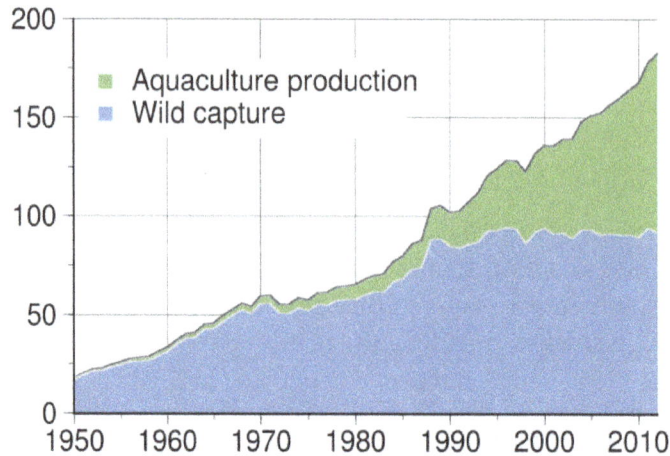

Global harvest of aquatic organisms in million tonnes, 1950–2010, as reported by the FAO

Particular kinds of aquaculture include fish farming, shrimp farming, oyster farming, mariculture, algaculture (such as seaweed farming), and the cultivation of ornamental fish. Particular methods include aquaponics and integrated multi-trophic aquaculture, both of which integrate fish farming and plant farming.

## History

The indigenous Gunditjmara people in Victoria, Australia, may have raised eels as early as 6000 BC. Evidence indicates they developed about 100 km² (39 sq mi) of volcanic floodplains in the vicinity of Lake Condah into a complex of channels and dams, and used woven traps to capture eels, and preserve them to eat all year round.

Workers harvest catfish from the Delta Pride Catfish farms in Mississippi

Aquaculture was operating in China *circa* 2500 BC. When the waters subsided after river floods, some fish, mainly carp, were trapped in lakes. Early aquaculturists fed their brood using nymphs and silkworm feces, and ate them. A fortunate genetic mutation of carp led to the emergence of goldfish during the Tang dynasty.

Japanese cultivated seaweed by providing bamboo poles and, later, nets and oyster shells to serve as anchoring surfaces for spores.

Romans bred fish in ponds and farmed oysters in coastal lagoons before 100 CE.

In central Europe, early Christian monasteries adopted Roman aquacultural practices. Aquaculture spread in Europe during the Middle Ages since away from the seacoasts and the big rivers, fish had to be salted so they did not rot. Improvements in transportation during the 19th century made fresh fish easily available and inexpensive, even in inland areas, making aquaculture less popular. The 15th-century fishponds of the Trebon Basin in the Czech Republic are maintained as a UNESCO World Heritage Site.

Hawaiians constructed oceanic fish ponds. A remarkable example is a fish pond dating from at least 1,000 years ago, at Alekoko. Legend says that it was constructed by the mythical Menehune dwarf people.

In first half of 18th century, German Stephan Ludwig Jacobi experimented with external fertilization of brown trouts and salmon. He wrote an article *"Von der künstlichen Erzeugung der Forellen und Lachse"*. By the latter decades of the 18th century, oyster farming had begun in estuaries along the Atlantic Coast of North America.

The word aquaculture appeared in an 1855 newspaper article in reference to the harvesting of ice. It also appeared in descriptions of the terrestrial agricultural practise of subirrigation in the late 19th century before becoming associated primarily with the cultivation of aquatic plant and animal species.

In 1859, Stephen Ainsworth of West Bloomfield, New York, began experiments with brook trout. By 1864, Seth Green had established a commercial fish-hatching operation at Caledonia Springs, near Rochester, New York. By 1866, with the involvement of Dr. W. W. Fletcher of Concord, Massachusetts, artificial fish hatcheries were under way in both Canada and the United States. When the Dildo Island fish hatchery opened in Newfoundland in 1889, it was the largest and most advanced in the world. The word aquaculture was used in descriptions of the hatcheries experiments with cod and lobster in 1890.

By the 1920s, the American Fish Culture Company of Carolina, Rhode Island, founded in the 1870s was one of the leading producers of trout. During the 1940s, they had perfected the method of manipulating the day and night cycle of fish so that they could be artificially spawned year around.

Californians harvested wild kelp and attempted to manage supply around 1900, later labeling it a wartime resource.

## 21st-century Practice

Harvest stagnation in wild fisheries and overexploitation of popular marine species, combined with a growing demand for high-quality protein, encouraged aquaculturists to domesticate other marine species. At the outset of modern aquaculture, many were optimistic that a "Blue Revolution" could take place in aquaculture, just as the Green Revolution of the 20th century had revolutionized agriculture. Although land animals had long been domesticated, most seafood species were still caught from the wild. Concerned about the impact of growing demand for seafood on the world's oceans, prominent ocean explorer Jacques Cousteau wrote in 1973: "With earth's burgeoning human populations to feed, we must turn to the sea with new understanding and new technology."

About 430 (97%) of the species cultured as of 2007 were domesticated during the 20th and 21st centuries, of which an estimated 106 came in the decade to 2007. Given the long-term importance of agriculture, to date, only 0.08% of known land plant species and 0.0002% of known land animal species have been domesticated, compared with 0.17% of known marine plant species and 0.13% of known marine animal species. Domestication typically involves about a decade of scientific research. Domesticating aquatic species involves fewer risks to humans than do land animals, which took a large toll in human lives. Most major human diseases originated in domesticated animals, including diseases such as smallpox and diphtheria, that like most infectious diseases, move to humans from animals. No human pathogens of comparable virulence have yet emerged from marine species.

Biological control methods to manage parasites are already being used, such as cleaner fish (e.g. lumpsuckers and wrasse) to control sea lice populations in salmon farming. Models are being used to help with spatial planning and siting of fish farms in order to minimize impact.

The decline in wild fish stocks has increased the demand for farmed fish. However, finding alternative sources of protein and oil for fish feed is necessary so the aquaculture industry can grow sustainably; otherwise, it represents a great risk for the over-exploitation of forage fish.

Another recent issue following the banning in 2008 of organotins by the International Maritime Organization is the need to find environmentally friendly, but still effective, compounds with antifouling effects.

Many new natural compounds are discovered every year, but producing them on a large enough scale for commercial purposes is almost impossible.

It is highly probable that future developments in this field will rely on microorganisms, but greater funding and further research is needed to overcome the lack of knowledge in this field.

**Species Groups**

**Global Aquaculture Production in Million Tonnes, 1950–2010, as Reported by the FAO**

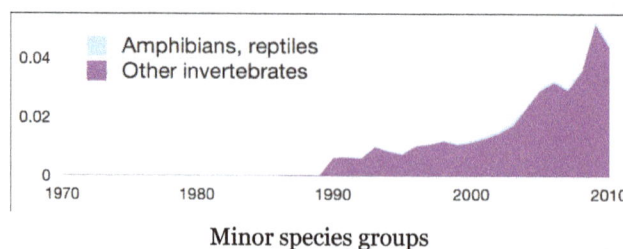

Minor species groups

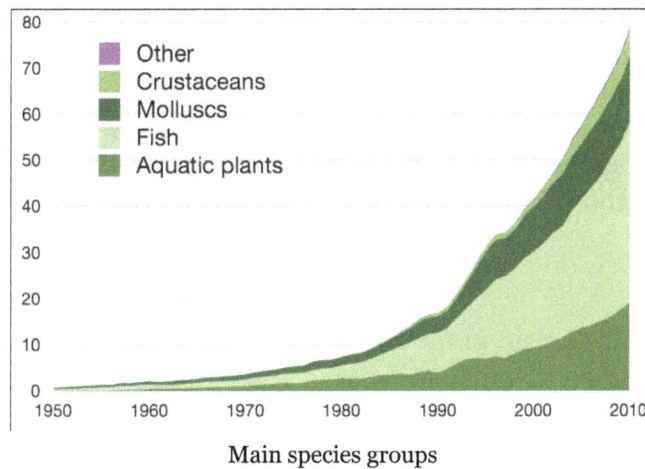

Main species groups

## Aquatic Plants

Cultivating emergent aquatic plants in floating containers

Microalgae, also referred to as phytoplankton, microphytes, or planktonic algae, constitute the majority of cultivated algae. Macroalgae commonly known as seaweed also have many commercial and industrial uses, but due to their size and specific requirements, they are not easily cultivated on a large scale and are most often taken in the wild.

## Fish

The farming of fish is the most common form of aquaculture. It involves raising fish commercially in tanks, ponds, or ocean enclosures, usually for food. A facility that releases juvenile fish into the wild for recreational fishing or to supplement a species' natural numbers is generally referred to as a fish hatchery. Worldwide, the most important fish species used in fish farming are, in order, carp, salmon, tilapia, and catfish.

In the Mediterranean, young bluefin tuna are netted at sea and towed slowly towards the shore. They are then interned in offshore pens where they are further grown for the market. In 2009, researchers in Australia managed for the first time to coax southern bluefin tuna to breed in land-locked tanks. Southern bluefin tuna are also caught in the wild and fattened in grow-out sea cages in southern Spencer Gulf, South Australia.

A similar process is used in the salmon-farming section of this industry; juveniles are taken from hatcheries and a variety of methods are used to aid them in their maturation. For example, as stated above, some of the most important fish species in the industry, salmon, can be grown using a cage system. This is done by having netted cages, preferably in open water that has a strong flow, and feeding the salmon a special food mixture that aids their growth. This process allows for year-round growth of the fish, thus a higher harvest during the correct seasons.

## Crustaceans

Commercial shrimp farming began in the 1970s, and production grew steeply thereafter. Global production reached more than 1.6 million tonnes in 2003, worth about US$9 billion. About 75% of farmed shrimp is produced in Asia, in particular in China and Thailand. The other 25% is produced mainly in Latin America, where Brazil is the largest producer. Thailand is the largest exporter.

Shrimp farming has changed from its traditional, small-scale form in Southeast Asia into a global industry. Technological advances have led to ever higher densities per unit area, and broodstock is shipped worldwide. Virtually all farmed shrimp are penaeids (i.e., shrimp of the family Penaeidae), and just two species of shrimp, the Pacific white shrimp and the giant tiger prawn, account for about 80% of all farmed shrimp. These industrial monocultures are very susceptible to disease, which has decimated shrimp populations across entire regions. Increasing ecological problems, repeated disease outbreaks, and pressure and criticism from both nongovernmental organizations and consumer countries led to changes in the industry in the late 1990s and generally stronger regulations. In 1999, governments, industry representatives, and environmental organizations initiated a program aimed at developing and promoting more sustainable farming practices through the Seafood Watch program.

Freshwater prawn farming shares many characteristics with, including many problems with, marine shrimp farming. Unique problems are introduced by the developmental lifecycle of the main species, the giant river prawn.

The global annual production of freshwater prawns (excluding crayfish and crabs) in 2003 was about 280,000 tonnes, of which China produced 180,000 tonnes followed by India and Thailand with 35,000 tonnes each. Additionally, China produced about 370,000 tonnes of Chinese river crab.

## Molluscs

Abalone farm

Aquacultured shellfish include various oyster, mussel, and clam species. These bivalves are filter and/or deposit feeders, which rely on ambient primary production rather than inputs of fish or other feed. As such, shellfish aquaculture is generally perceived as benign or even beneficial.

Depending on the species and local conditions, bivalve molluscs are either grown on the beach, on longlines, or suspended from rafts and harvested by hand or by dredging.

Abalone farming began in the late 1950s and early 1960s in Japan and China. Since the mid-1990s, this industry has become increasingly successful. Overfishing and poaching have reduced wild populations to the extent that farmed abalone now supplies most abalone meat. Sustainably farmed molluscs can be certified by Seafood Watch and other organizations, including the World Wildlife Fund (WWF). WWF initiated the "Aquaculture Dialogues" in 2004 to develop measurable and performance-based standards for responsibly farmed seafood. In 2009, WWF co-founded the Aquaculture Stewardship Council with the Dutch Sustainable Trade Initiative to manage the global standards and certification programs.

After trials in 2012, a commercial "sea ranch" was set up in Flinders Bay, Western Australia, to raise abalone. The ranch is based on an artificial reef made up of 5000 (As of April 2016[update]) separate concrete units called 'abitats' (abalone habitats). The 900-kg abitats can host 400 abalone each. The reef is seeded with young abalone from an onshore hatchery. The abalone feed on seaweed that has grown naturally on the abitats, with the ecosystem enrichment of the bay also resulting in growing numbers of dhufish, pink snapper, wrasse, and Samson fish, among other species.

Brad Adams, from the company, has emphasised the similarity to wild abalone and the difference from shore-based aquaculture. "We're not aquaculture, we're ranching, because once they're in the water they look after themselves."

## Other Groups

Other groups include aquatic reptiles, amphibians, and miscellaneous invertebrates, such as echinoderms and jellyfish. They are separately graphed at the top right of this section, since they do not contribute enough volume to show clearly on the main graph.

Commercially harvested echinoderms include sea cucumbers and sea urchins. In China, sea cucumbers are farmed in artificial ponds as large as 1,000 acres (400 ha).

## Around the World

### Global Aquaculture Production in Million Tonnes, 1950–2010, as Reported by the FAO

In 2012, the total world production of fisheries was 158 million tonnes, of which aquaculture contributed 66.6 million tonnes, about 42%. The growth rate of worldwide aquaculture has been sustained and rapid, averaging about 8% per year for over 30 years, while the take from wild fisheries] has been essentially flat for the last decade. The aquaculture market reached $86 billion in 2009.

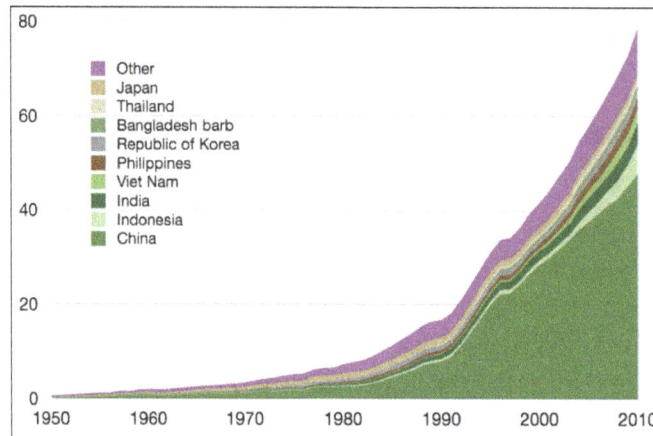

Main aquaculture countries, 1950–2010

Aquaculture is an especially important economic activity in China. Between 1980 and 1997, the Chinese Bureau of Fisheries reports, aquaculture harvests grew at an annual rate of 16.7%, jumping from 1.9 million tonnes to nearly 23 million tonnes. In 2005, China accounted for 70% of world production. Aquaculture is also currently one of the fastest-growing areas of food production in the U.S.

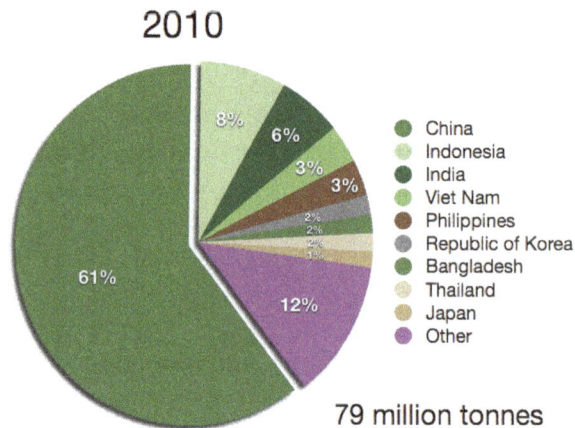

Main aquaculture countries in 2010

About 90% of all U.S. shrimp consumption is farmed and imported. In recent years, salmon aquaculture has become a major export in southern Chile, especially in Puerto Montt, Chile's fastest-growing city.

A United Nations report titled *The State of the World Fisheries and Aquaculture* released in May 2014 maintained fisheries and aquaculture support the livelihoods of some 60 million people in Asia and Africa.

## National Laws, Regulations, and Management

Laws governing aquaculture practices vary greatly by country and are often not closely regulated or easily traceable. In the United States, land-based and nearshore aquaculture is regulated at the federal and state levels; however, no national laws govern offshore aquaculture in U.S. exclusive economic zone waters. In June 2011, the Department of Commerce and National Oceanic and Atmospheric Administration released national aquaculture policies to address this issue and "to meet the growing demand for healthy seafood, to create jobs in coastal communities, and restore vital ecosystems." In 2011, Congresswoman Lois Capps introduced the *National Sustainable Offshore Aquaculture Act of 2011* "to establish a regulatory system and research program for sustainable offshore aquaculture in the United States exclusive economic zone"; however, the bill was not enacted into law.

## Over-reporting

China overwhelmingly dominates the world in reported aquaculture output, reporting a total output which is double that of the rest of the world put together. However, issues exist with the accuracy of China's returns.

In 2001, fisheries scientists Reg Watson and Daniel Pauly expressed concerns in a letter to *Nature*, that China was over-reporting its catch from wild fisheries in the 1990s. They said that made it appear that the global catch since 1988 was increasing annually by 300,000 tonnes, whereas it was really shrinking annually by 350,000 tonnes. Watson and Pauly suggested this may be related to China policies where state entities that monitor the economy are also asked to increase output. Also, until recently, the promotion of Chinese officials was based on production increases from their own areas.

China disputes this claim. The official Xinhua News Agency quoted Yang Jian, director general of the Agriculture Ministry's Bureau of Fisheries, as saying that China's figures were "basically correct". However, the FAO accepts issues exist with the reliability of China's statistical returns, and currently treats data from China, including the aquaculture data, apart from the rest of the world.

## Aquacultural Methods

## Mariculture

Mariculture refers to the cultivation of marine organisms in seawater, usually in sheltered coastal waters. The farming of marine fish is an example of mariculture, and so also is the farming of marine crustaceans (such as shrimp), molluscs (such as oysters), and seaweed.

Mariculture off High Island, Hong Kong

Mariculture may consist of raising the organisms on or in artificial enclosures such as in floating netted enclosures for salmon and on racks for oysters. In the case of enclosed salmon, they are fed by the operators; oysters on racks filter feed on naturally available food. Abalone have been farmed on an artificial reef consuming seaweed which grows naturally on the reef units.

The adaptable tilapia is another commonly farmed fish

Carp are the dominant fish in aquaculture

## Integrated

Integrated multi-trophic aquaculture (IMTA) is a practice in which the byproducts (wastes) from one species are recycled to become inputs (fertilizers, food) for another. Fed aquaculture (for example, fish, shrimp) is combined with inorganic extractive and organic extractive (for example, shellfish) aquaculture to create balanced systems for environmental sustainability (biomitigation), economic stability (product diversification and risk reduction) and social acceptability (better management practices).

"Multi-trophic" refers to the incorporation of species from different trophic or nutritional levels in the same system. This is one potential distinction from the age-old practice of aquatic polyculture, which could simply be the co-culture of different fish species from the same trophic level. In this case, these organisms may all share the same biological and chemical processes, with few synergistic benefits, which could potentially lead to significant shifts in the ecosystem. Some traditional polyculture systems may, in fact, incorporate a greater diversity of species, occupying several niches, as extensive cultures (low intensity, low management) within the same pond. The term "integrated" refers to the more intensive cultivation of the different species in proximity of each other, connected by nutrient and energy transfer through water.

Ideally, the biological and chemical processes in an IMTA system should balance. This is achieved through the appropriate selection and proportions of different species providing different ecosystem functions. The co-cultured species are typically more than just biofilters; they are harvestable crops of commercial value. A working IMTA system can result in greater total production based on mutual benefits to the co-cultured species and improved ecosystem health, even if the production of individual species is lower than in a monoculture over a short term period.

Sometimes the term "integrated aquaculture" is used to describe the integration of monocultures through water transfer. For all intents and purposes, however, the terms "IMTA" and "integrated aquaculture" differ only in their degree of descriptiveness. Aquaponics, fractionated aquaculture, integrated agriculture-aquaculture systems, integrated peri-urban-aquaculture systems, and integrated fisheries-aquaculture systems are other variations of the IMTA concept.

## Netting Materials

Various materials, including nylon, polyester, polypropylene, polyethylene, plastic-coated welded wire, rubber, patented rope products (Spectra, Thorn-D, Dyneema), galvanized steel and copper are used for netting in aquaculture fish enclosures around the world. All of these materials are selected for a variety of reasons, including design feasibility, material strength, cost, and corrosion resistance.

Recently, copper alloys have become important netting materials in aquaculture because they are antimicrobial (i.e., they destroy bacteria, viruses, fungi, algae, and other microbes) and they therefore prevent biofouling (i.e., the undesirable accumulation, adhesion, and growth of microorganisms, plants, algae, tubeworms, barnacles, mollusks, and other organisms). By inhibiting microbial growth, copper alloy aquaculture cages avoid costly net changes that are necessary with other materials. The resistance of organism growth on copper alloy nets also provides a cleaner and healthier environment for farmed fish to grow and thrive.

## Issues

Aquaculture can be more environmentally damaging than exploiting wild fisheries on a local area basis but has considerably less impact on the global environment on a per kg of production basis. Local concerns include waste handling, side-effects of antibiotics, competition between farmed and wild animals, and using other fish to feed more marketable carnivorous fish. However, research and commercial feed improvements during the 1990s and 2000s have lessened many of these concerns.

Aquaculture may contribute to propagation of invasive species. As the cases of Nile perch and Janitor fish show, this issue may be damaging to native fauna.

Fish waste is organic and composed of nutrients necessary in all components of aquatic food webs. In-ocean aquaculture often produces much higher than normal fish waste concentrations. The waste collects on the ocean bottom, damaging or eliminating bottom-dwelling life. Waste can also decrease dissolved oxygen levels in the water column, putting further pressure on wild animals.

An alternative model to food being added to the ecosystem, is the installation of artificial reef structures to increase the habitat niches available, without the need to add any more than ambient feed and nutrient. This has been used in the "ranching" of abalone in Western Australia.

## Fish Oils

Tilapia from aquaculture has been shown to contain more fat and a much higher ratio of omega-6 to omega-3 oils.

## Impacts on Wild Fish

Some carnivorous and omnivorous farmed fish species are fed wild forage fish. Although carnivorous farmed fish represented only 13 percent of aquaculture production by weight in 2000, they represented 34 percent of aquaculture production by value.

Farming of carnivorous species like salmon and shrimp leads to a high demand for forage fish to match the nutrition they get in the wild. Fish do not actually produce omega-3 fatty acids, but instead accumulate them from either consuming microalgae that produce these fatty acids, as is the case with forage fish like herring and sardines, or, as is the case with fatty predatory fish, like salmon, by eating prey fish that have accumulated omega-3 fatty acids from microalgae. To satisfy this requirement, more than 50 percent of the world fish oil production is fed to farmed salmon.

Farmed salmon consume more wild fish than they generate as a final product, although the efficiency of production is improving. To produce one pound of farmed salmon, products from several pounds of wild fish are fed to them - this can be described as the "fish-in-fish-out" (FIFO) ratio. In 1995, salmon had a FIFO ratio of 7.5 (meaning 7.5 pounds of wild fish feed were required to produce 1 pound of salmon); by 2006 the ratio had fallen to 4.9. Additionally, a growing share of fish oil and fishmeal come from residues (byproducts of fish processing), rather than dedicated whole fish. In 2012, 34 percent of fish oil and 28 percent of fishmeal came from residues. However, fishmeal and oil from residues instead of whole fish have a different composition with more ash and less protein, which may limit its potential use for aquaculture.

As the salmon farming industry expands, it requires more wild forage fish for feed, at a time when seventy five percent of the worlds monitored fisheries are already near to or have exceeded their maximum sustainable yield. The industrial scale extraction of wild forage fish for salmon farming then impacts the survivability of the wild predator fish who rely on them for food. An important step in reducing the impact of aquaculture on wild fish is shifting carnivorous species to plant-based feeds. Salmon feeds, for example, have gone from containing only fishmeal and oil to containing 40 percent plant protein. The USDA has also experimented with using grain-based feeds for farmed trout. When properly formulated (and often mixed with fishmeal or oil), plant-based feeds can provide proper nutrition and similar growth rates in carnivorous farmed fish.

Another impact aquaculture production can have on wild fish is the risk of fish escaping from coastal pens, where they can interbreed with their wild counterparts, diluting wild genetic stocks. Escaped fish can become invasive, out-competing native species.

## Coastal Ecosystems

Aquaculture is becoming a significant threat to coastal ecosystems. About 20 percent of mangrove forests have been destroyed since 1980, partly due to shrimp farming. An extended cost–benefit analysis of the total economic value of shrimp aquaculture built on mangrove ecosystems found that the external costs were much higher than the external benefits. Over four decades, 269,000 hectares (660,000 acres) of Indonesian mangroves have been converted to shrimp farms. Most of these farms are abandoned within a decade because of the toxin build-up and nutrient loss.

Salmon farms are typically sited in pristine coastal ecosystems which they then pollute. A farm with 200,000 salmon discharges more fecal waste than a city of 60,000 people. This waste is discharged directly into the surrounding aquatic environment, untreated, often containing antibiotics and pesticides." There is also an accumulation of heavy metals on the benthos (seafloor) near the salmon farms, particularly copper and zinc.

In 2016, mass fish kill events impacted salmon farmers along Chile's coast and the wider ecology. Increases in aquaculture production and its associated effluent were considered to be possible contributing factors to fish and molluscan mortality.

## Genetic Modification

A type of salmon called the AquAdvantage salmon has been genetically modified for faster growth, although it has not been approved for commercial use, due to controversy. The altered salmon incorporates a growth hormone from a Chinook salmon that allows it to reach full size in 16-28 months, instead of the normal 36 months for Atlantic salmon, and while consuming 25 percent less feed. The U.S. Food and Drug Administration reviewed the AquAdvantage salmon in a draft environmental assessment and determined that it "would not have a significant impact (FONSI) on the U.S. environment."

## Animal Welfare

As with the farming of terrestrial animals, social attitudes influence the need for humane practices and regulations in farmed marine animals. Under the guidelines advised by the Farm Animal Welfare Council good animal welfare means both fitness and a sense of well being in the animal's physical and mental state. This can be defined by the Five Freedoms:

- Freedom from hunger & thirst

- Freedom from discomfort

- Freedom from pain, disease, or injury

- Freedom to express normal behaviour

- Freedom from fear and distress

However, the controversial issue in aquaculture is whether fish and farmed marine invertebrates are actually sentient, or have the perception and awareness to experience suffering. Although no evidence of this has been found in marine invertebrates, recent studies conclude that fish do have the necessary receptors (nociceptors) to sense noxious stimuli and so are likely to experience states of pain, fear and stress. Consequently, welfare in aquaculture is directed at vertebrates; finfish in particular.

## Common Welfare Concerns

Welfare in aquaculture can be impacted by a number of issues such as stocking densities, behavioural interactions, disease and parasitism. A major problem in determining the cause of impaired welfare is that these issues are often all interrelated and influence each other at different times.

Optimal stocking density is often defined by the carrying capacity of the stocked environment and the amount of individual space needed by the fish, which is very species specific. Although behavioural interactions such as shoaling may mean that high stocking densities are beneficial to some species, in many cultured species high stocking densities may be of concern. Crowding can constrain normal swimming behaviour, as well as increase aggressive and competitive behaviours such as cannibalism, feed competition, territoriality and dominance/subordination hierarchies. This potentially increases the risk of tissue damage due to abrasion from fish-to-fish contact or fish-to-cage contact. Fish can suffer reductions in food intake and food conversion efficiency. In addition, high stocking densities can result in water flow being insufficient, creating inadequate oxygen supply and waste product removal. Dissolved oxygen is essential for fish respiration and concentrations below critical levels can induce stress and even lead to asphyxiation. Ammonia, a nitrogen excretion product, is highly toxic to fish at accumulated levels, particularly when oxygen concentrations are low.

Many of these interactions and effects cause stress in the fish, which can be a major factor in facilitating fish disease. For many parasites, infestation depends on the host's degree of mobility, the density of the host population and vulnerability of the host's defence system. Sea lice are the primary parasitic problem for finfish in aquaculture, high numbers causing widespread skin erosion and haemorrhaging, gill congestion,and increased mucus production. There are also a number of prominent viral and bacterial pathogens that can have severe effects on internal organs and nervous systems.

## Improving Welfare

The key to improving welfare of marine cultured organisms is to reduce stress to a minimum, as prolonged or repeated stress can cause a range of adverse effects. Attempts to minimise stress can occur throughout the culture process. During grow out it is important to keep stocking densities at appropriate levels specific to each species, as well as separating size classes and grading to reduce aggressive behavioural interactions. Keeping nets and cages clean can assist positive water flow to reduce the risk of water degradation.

Not surprisingly disease and parasitism can have a major effect on fish welfare and it is important for farmers not only to manage infected stock but also to apply disease prevention measures. However, prevention methods, such as vaccination, can also induce stress because of the extra handling and injection. Other methods include adding antibiotics to feed, adding chemicals into water for treatment baths and biological control, such as using cleaner wrasse to remove lice from farmed salmon.

Many steps are involved in transport, including capture, food deprivation to reduce faecal contamination of transport water, transfer to transport vehicle via nets or pumps, plus transport and transfer to the delivery location. During transport water needs to be maintained to a high quality, with regulated temperature, sufficient oxygen and minimal waste products. In some cases anaesthetics may be used in small doses to calm fish before transport.

Aquaculture is sometimes part of an environmental rehabilitation program or as an aid in conserving endangered species.

## Prospects

Global wild fisheries are in decline, with valuable habitat such as estuaries in critical condition. The aquaculture or farming of piscivorous fish, like salmon, does not help the problem because they need to eat products from other fish, such as fish meal and fish oil. Studies have shown that salmon farming has major negative impacts on wild salmon, as well as the forage fish that need to be caught to feed them. Fish that are higher on the food chain are less efficient sources of food energy.

Apart from fish and shrimp, some aquaculture undertakings, such as seaweed and filter-feeding bivalve mollusks like oysters, clams, mussels and scallops, are relatively benign and even environmentally restorative. Filter-feeders filter pollutants as well as nutrients from the water, improving water quality. Seaweeds extract nutrients such as inorganic nitrogen and phosphorus directly from the water, and filter-feeding mollusks can extract nutrients as they feed on particulates, such as phytoplankton and detritus.

Some profitable aquaculture cooperatives promote sustainable practices. New methods lessen the risk of biological and chemical pollution through minimizing fish stress, fallowing netpens, and applying Integrated Pest Management. Vaccines are being used more and more to reduce antibiotic use for disease control.

Onshore recirculating aquaculture systems, facilities using polyculture techniques, and properly sited facilities (for example, offshore areas with strong currents) are examples of ways to manage negative environmental effects.

Recirculating aquaculture systems (RAS) recycle water by circulating it through filters to remove fish waste and food and then recirculating it back into the tanks. This saves water and the waste gathered can be used in compost or, in some cases, could even be treated and used on land. While RAS was developed with freshwater fish in mind, scientist associated with the Agricultural Research Service have found a way to rear saltwater fish using RAS in low-salinity waters. Although saltwater fish are raised in off-shore cages or caught with nets in water that typically has a salinity of 35 parts per thousand (ppt), scientists were able to produce healthy pompano, a saltwater fish, in tanks with a salinity of only 5 ppt. Commercializing low-salinity RAS are predicted to have positive environmental and economical effects. Unwanted nutrients from the fish food would not be added to the ocean and the risk of transmitting diseases between wild and farm-raised fish would greatly be reduced. The price of expensive saltwater fish, such as the pompano and combia used in the experiments, would be reduced. However, before any of this can be done researchers must study every aspect of the fish's lifecycle, including the amount of ammonia and nitrate the fish will tolerate in the water, what to feed the fish during each stage of its lifecycle, the stocking rate that will produce the healthiest fish, etc.

Some 16 countries now use geothermal energy for aquaculture, including China, Israel, and the United States. In California, for example, 15 fish farms produce tilapia, bass, and catfish with warm water from underground. This warmer water enables fish to grow all year round and mature more quickly. Collectively these California farms produce 4.5 million kilograms of fish each year.

## References

- GESAMP (2008) Assessment and communication of environmental risks in coastal aquaculture FAO Reports and Studies No 76. ISBN 978-92-5-105947-0

- Wyban, Carol Araki (1992) Tide and Current: Fishponds of Hawai'I University of Hawaii Press:: ISBN 978-0-8248-1396-3

# 2

# Various Branches of Aquaculture

Aquaculture has proven to be a very popular food system due to its commercial viability, potential for entrepreneurial development and job creating capability. It can also be adapted to any region that can support fisheries. The major components of aquaculture such as mariculture and oyster farming are discussed in this chapter.

## Mariculture

Mariculture is a specialized branch of aquaculture involving the cultivation of marine organisms for food and other products in the open ocean, an enclosed section of the ocean, or in tanks, ponds or raceways which are filled with seawater. An example of the latter is the farming of marine fish, including finfish and shellfish like prawns, or oysters and seaweed in saltwater ponds. Non-food products produced by mariculture include: fish meal, nutrient agar, jewellery (e.g. cultured pearls), and cosmetics.

Fish cages containing salmon in Loch Ailort, Scotland.

### Methods

### Shellfish

Similar to algae cultivation, shellfish can be farmed in multiple ways: on ropes, in bags or cages, or directly on (or within) the intertidal substrate. Shellfish mariculture does not require feed or fertilizer inputs, nor insecticides or antibiotics, making shellfish aquaculture (or 'mariculture') a self-supporting system. Shellfish can also be used in multi-species cultivation techniques, where shellfish can utilize waste generated by higher trophic level organisms.

## Artificial Reefs

After trials in 2012, a commercial "sea ranch" was set up in Flinders Bay, Western Australia to raise abalone. The ranch is based on an artificial reef made up of 5000 (As of April 2016) separate concrete units called *abitats* (abalone habitats). The 900 kilograms (2,000 lb) abitats can host 400 abalone each. The reef is seeded with young abalone from an onshore hatchery.

The abalone feed on seaweed that has grown naturally on the abitats; with the ecosystem enrichment of the bay also resulting in growing numbers of dhufish, pink snapper, wrasse, Samson fish among other species.

Brad Adams, from the company, has emphasised the similarity to wild abalone and the difference from shore based aquaculture. "We're not aquaculture, we're ranching, because once they're in the water they look after themselves."

## Open Ocean

Raising marine organisms under controlled conditions in exposed, high-energy ocean environments beyond significant coastal influence, is a relatively new approach to mariculture. Open ocean aquaculture (OOA) uses cages, nets, or long-line arrays that are moored, towed or float freely. Research and commercial open ocean aquaculture facilities are in operation or under development in Panama, Australia, Chile, China, France, Ireland, Italy, Japan, Mexico, and Norway. As of 2004, two commercial open ocean facilities were operating in U.S. waters, raising Threadfin near Hawaii and cobia near Puerto Rico. An operation targeting bigeye tuna recently received final approval. All U.S. commercial facilities are currently sited in waters under state or territorial jurisdiction. The largest deep water open ocean farm in the world is raising cobia 12 km off the northern coast of Panama in highly exposed sites.

## Enhanced Stocking

Enchanced Stocking (also known as sea ranching) is a Japanese principle based on operant conditioning and the migratory nature of certain species. The fishermen raise hatchlings in a closely knitted net in a harbor, sounding an underwater horn before each feeding. When the fish are old enough they are freed from the net to mature in the open sea. During spawning season, about 80% of these fish return to their birthplace. The fishermen sound the horn and then net those fish that respond.

## Seawater Ponds

In seawater pond mariculture, fish are raised in ponds which receive water from the sea. This has the benefit that the nutrition (e.g. microorganisms) present in the seawater can be used. This is a great advantage over traditional fish farms (e.g. sweet water farms) for which the farmers buy feed (which is expensive). Other advantages are that water purification plants may be planted in the ponds to eliminate the buildup of nitrogen, from fecal and other contamination. Also, the ponds can be left unprotected from natural predators, providing another kind of filtering.

## Environmental Effects

Mariculture has rapidly expanded over the last two decades due to new technology, improvements in formulated feeds, greater biological understanding of farmed species, increased water quality within closed farm systems, greater demand for seafood products, site expansion and government interest. As a consequence, mariculture has been subject to some controversy regarding its social and environmental impacts. Commonly identified environmental impacts from marine farms are:

1. Wastes from cage cultures;

2. Farm escapees and invasives;

3. Genetic pollution and disease and parasite transfer;

4. Habitat modification.

As with most farming practices, the degree of environmental impact depends on the size of the farm, the cultured species, stock density, type of feed, hydrography of the site, and husbandry methods. The adjacent diagram connects these causes and effects.

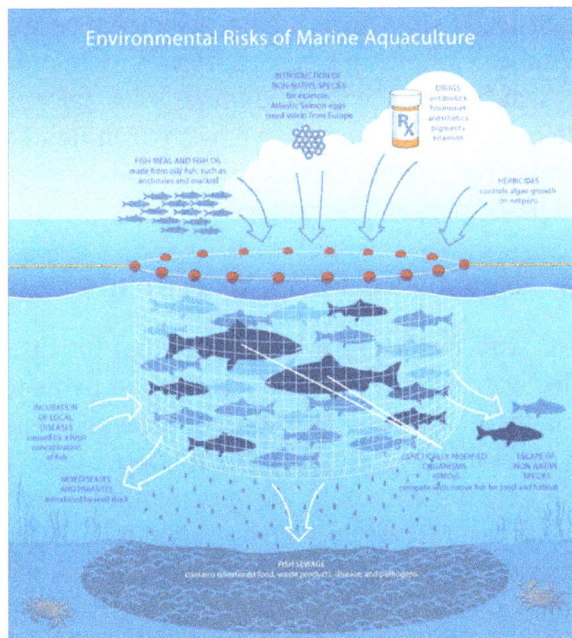

## Wastes from Cage Cultures

Mariculture of finfish can require a significant amount of fishmeal or other high protein food sources. Originally, a lot of fishmeal went to waste due to inefficient feeding regimes and poor digestibility of formulated feeds which resulted in poor feed conversion ratios.

In cage culture, several different methods are used for feeding farmed fish – from simple hand feeding to sophisticated computer-controlled systems with automated food dispensers coupled with *in situ* uptake sensors that detect consumption rates. In coastal fish farms, overfeeding primarily leads to increased disposition of detritus on the seafloor (potentially smothering seafloor dwelling invertebrates and altering the physical environment), while in hatcheries and land-based farms, excess food goes to waste and can potentially impact the surrounding catchment and local coastal environment. This impact is usually highly local, and depends significantly on the settling velocity of waste feed and the current velocity (which varies both spatially and temporally) and depth.

## Farm Escapees and Invasives

The impact of escapees from aquaculture operations depends on whether or not there are wild conspecifics or close relatives in the receiving environment, and whether or not the escapee is reproductively capable. Several different mitigation/prevention strategies are currently employed, from the development of infertile triploids to land-based farms which are completely isolated from any marine environment. Escapees can adversely impact local ecosystems through hybridization and loss of genetic diversity in native stocks, increase negative interactions within an ecosystem (such as predation and competition), disease transmission and habitat changes (from trophic cascades and ecosystem shifts to varying sediment regimes and thus turbidity).

The accidental introduction of invasive species is also of concern. Aquaculture is one of the main vectors for invasives following accidental releases of farmed stocks into the wild. One example is the Siberian sturgeon (*Acipenser baerii*) which accidentally escaped from a fish farm into the Gironde Estuary (Southwest France) following a severe storm in December 1999 (5,000 individual fish escaped into the estuary which had never hosted this species before). Molluscan farming is another example whereby species can be introduced to new environments by 'hitchhiking' on farmed molluscs. Also, farmed molluscs themselves can become dominate predators and/or competitors, as well as potentially spread pathogens and parasites.

## Genetic Pollution and Disease and Parasite Transfer

One of the primary concerns with mariculture is the potential for disease and parasite transfer. Farmed stocks are often selectively bred to increase disease and parasite resistance, as well as improving growth rates and quality of products. As a consequence, the genetic diversity within reared stocks decreases with every generation – meaning they can potentially reduce the genetic diversity within wild populations if they escape into those wild populations. Such genetic pollution from escaped aquaculture stock can reduce the wild population's ability to adjust to the changing natural environment. Also, maricultured species can harbour diseases and parasites (e.g., lice) which can be introduced to wild populations upon their escape. An example of this is the parasitic sea lice on wild and farmed Atlantic salmon in Canada. Also, non-indigenous species which are farmed may have resistance to, or carry, particular diseases (which they picked up in their native habitats) which could be spread through wild populations if they escape into those wild populations. Such 'new' diseases would be devastating for those wild populations because they would have no immunity to them.

## Habitat Modification

With the exception of benthic habitats directly beneath marine farms, most mariculture causes minimal destruction to habitats. However, the destruction of mangrove forests from the farming of shrimps is of concern. Globally, shrimp farming activity is a small contributor to the destruction of mangrove forests; however, locally it can be devastating. Mangrove forests provide rich matrices which support a great deal of biodiversity – predominately juvenile fish and crustaceans. Furthermore, they act as buffering systems whereby they reduce coastal erosion, and improve water quality for in situ animals by processing material and 'filtering' sediments.

## Others

In addition, nitrogen and phosphorus compounds from food and waste may lead to blooms of phytoplankton, whose subsequent degradation can drastically reduce oxygen levels. If the algae are toxic, fish are killed and shellfish contaminated.

## Sustainability

Mariculture development must be sustained by basic and applied research and development in major fields such as nutrition, genetics, system management, product handling, and socioeconomics. One approach is closed systems that have no direct interaction with the local environment. However, investment and operational cost are currently significantly higher than open cages, limiting them to their current role as hatcheries.

## Benefits

Sustainable mariculture promises economic and environmental benefits. Economies of scale imply that ranching can produce fish at lower cost than industrial fishing, leading to better human diets and the gradual elimination of unsustainable fisheries. Maricultured fish are also perceived to be of higher quality than fish raised in ponds or tanks, and offer more diverse choice of species. Consistent supply and quality control has enabled integration in food market channels.

## Species Farmed

Fish

- Seabass
- Bigeye tuna
- Cobia
- Grouper
- Snapper
- Pompano
- Salmon
- Pearlspot
- Mullet
- Pomfret

Shellfish/Crustaceans

- Abalone

- Oysters

- Prawn

- Mussels

Plants

- Seaweeds

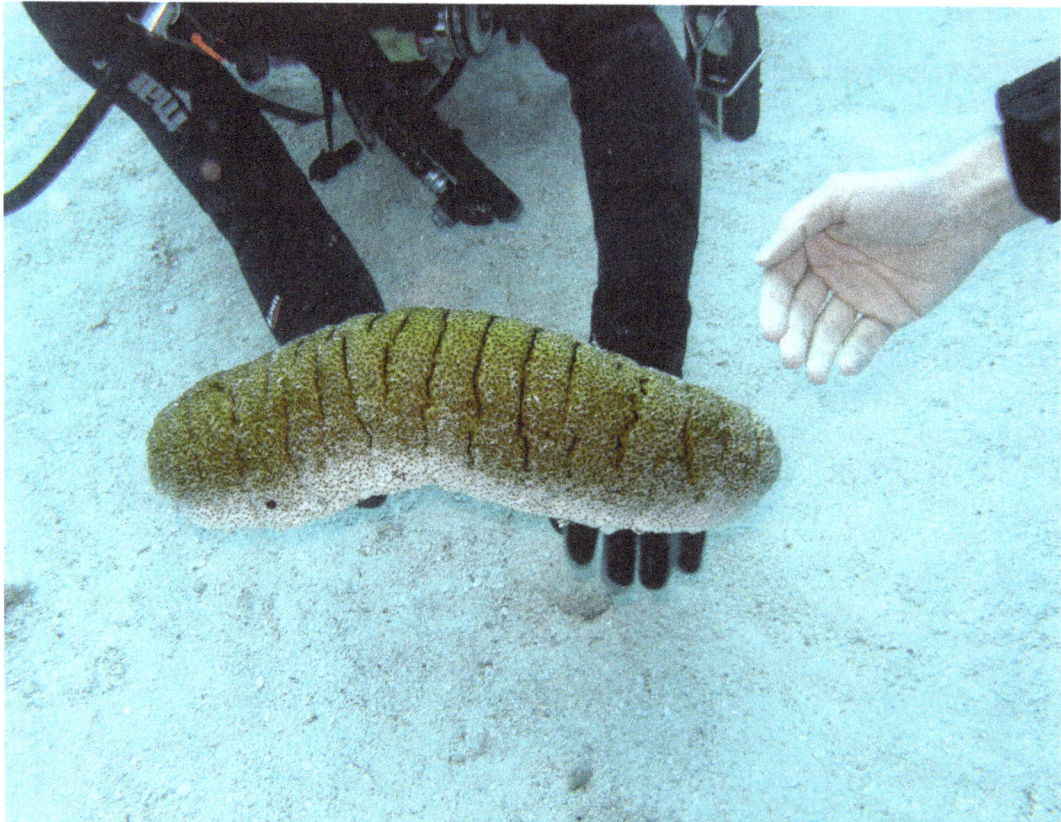

## Energing Approach to Mariculture

## Offshore Aquaculture

Offshore aquaculture, also known as open ocean aquaculture, is an emerging approach to mariculture or marine farming where fish farms are moved some distance offshore. The farms are positioned in deeper and less sheltered waters, where ocean currents are stronger than they are inshore.

One of the concerns with inshore aquaculture is that discarded nutrients and feces can settle below the farm on the seafloor and damage the benthic ecosystem. According to its proponents, the wastes from aquaculture that has been moved offshore tend to be swept away from the site and diluted. Moving aquaculture offshore also provides more space where aquaculture production can expand to meet the increasing demands for fish. It avoids many of the conflicts that occur with other marine resource users in the more crowded inshore waters, though there can still be user conflicts offshore. Critics are concerned about issues such as the ongoing consequences of using antibiotics and other drugs and the possibilities of cultured fish escaping and spreading disease among wild fish.

Offshore aquaculture uses fish cages similar to these inshore ones, except they are submerged and moved offshore into deeper water.

## Background

Aquaculture is the most rapidly expanding food industry in the world as a result of declining wild fisheries stocks and profitable business. In 2008, aquaculture provided 45.7% of the fish produced globally for human consumption; increasing at a mean rate of 6.6% a year since 1970.

In 1970, a National Oceanic and Atmospheric Administration (NOAA) grant brought together a group of oceanographers, engineers and marine biologists to explore whether offshore aquaculture, which was then considered a futuristic activity, was feasible. In the United States, the future of offshore aquaculture technology within federal waters has become much talked-about. As many commercial operations show, it is now technically possible to culture finfish, shellfish, and seaweeds using offshore aquaculture technology.

Major challenges for the offshore aquaculture industry involve designing and deploying cages that can withstand storms, dealing with the logistics of working many kilometers from land, and finding species that are sufficiently profitable to cover the costs of rearing fish in exposed offshore areas.

## Technology

To withstand the high energy offshore environment, farms must be built to be more robust than those inshore. However, the design of the offshore technology is developing rapidly, aimed at reducing cost and maintenance.

While the ranching systems currently used for tuna use open net cages at the surface of the sea (as is done also in salmon farming), the offshore technology usually uses submersible cages. These large rigid cages – each one able to hold many thousands of fish – are anchored on the sea floor, but can move up and down the water column. They are attached to buoys on the surface which frequently contain a mechanism for feeding and storage for equipment. Similar technology is being used in waters near the Bahamas, China, the Philippines, Portugal, Puerto Rico, and Spain. By submerging cages or shellfish culture systems, wave effects are minimized and interference with boating and shipping is reduced. Offshore farms can be made more efficient and safer if remote control is used, and technologies such as an 18-tonne buoy that feeds and monitors fish automatically over long periods are being developed.

## Existing Offshore Structures

Multi-functional use of offshore waters can lead to more sustainable aquaculture "in areas that can be simultaneously used for other activities such as energy production". Operations for finfish and shellfish are being developed. For example, the Hubb-Sea World Research Institutes' project to convert a retired oil platform 10 nm off the southern California coast to an experimental offshore aquaculture facility. The institute plans to grow mussels and red abalone on the actual platform, as well as white seabass, striped bass, bluefin tuna, California halibut and California yellowtail in floating cages.

## Integrated Multi-trophic Aquaculture

Integrated multi-trophic aquaculture (IMTA), or polyculture, occurs when species which must be fed, such as finfish, are cultured alongside species which can feed on dissolved nutrients, such as seaweeds, or organic wastes, such as suspension feeders and deposit feeders. This sustainable method could solve several problems with offshore aquaculture. The method is being pioneered in Spain, Canada, and elsewhere.

## Roaming Cages

Roaming cages have been envisioned as the "next generation technology" for offshore aquaculture. These are large mobile cages powered by thrusters and able to take advantage of ocean currents. One idea is that juvenile tuna, starting out in mobile cages in Mexico, could reach Japan after a few months, matured and ready for the market. However, implementing such ideas will have regulatory and legal implications.

## Space Conflicts

As oceans industrialise, conflicts are increasing among the users of marine space. This competition for marine space is developing in a context where natural resources can be seen as publicly owned. There can be conflict with the tourism industry, recreational fishers, wild harvest fisheries and the siting of marine renewable energy installations. The problems can be aggravated by the remoteness of many marine areas, and difficulties with monitoring and enforcement. On the other hand, remote sites can be chosen that avoid conflicts with other users, and allow large scale operations with resulting economies of scale. Offshore systems can provide alternatives for countries with few suitable inshore sites, like Spain.

## Ecological Impacts

The ecological impacts of offshore aquaculture are somewhat uncertain because it is still largely in the research stage. Many of the concerns over potential offshore aquaculture impacts are paralleled by similar, well established concerns over inshore aquaculture practices.

# Pollution

One of the concerns with inshore farms is that discarded nutrients and feces can settle on the seafloor and disturb the benthos. The "dilution of nutrients" that occurs in deeper water is a strong reason to move coastal aquaculture offshore into the open ocean. How much nutrient pollution and damage to the seafloor occurs depends on the feed conversion efficiency of the species, the flushing rate and the size of the operation. However, dissolved and particulate nutrients are still released to the environment. Future offshore farms will probably be much larger than inshore farms today, and will therefore generate more waste. The point at which the capacity of offshore ecosystems to assimilate waste from offshore aquaculture operations will be exceeded is yet to be defined.

# Wild Caught Feed

As with the inshore aquaculture of carnivorous fish, a large proportion of the feed comes from wild forage fish. Except for a few countries, offshore aquaculture has focused predominantly on high value carnivorous fish. If the industry attempts to expand with this focus then the supply of these wild fish will become ecologically unsustainable.

# Fish Escapes

The expense of offshore systems means it is important to avoid fish escapes. However, it is likely there will be escapes as the offshore industry expands. This could have significant consequences for native species, even if the farmed fish are inside their native range. Submersible cages are fully closed and therefore escapes can only occur through damage to the structure. Offshore cages must withstand the high energy of the environment and attacks by predators such as sharks. The outer netting is made of Spectra – a super-strong polyethylene fibre – wrapped tightly around the frame, leaving no slack for predators to grip. However, the fertilised eggs of cod are able to pass through the cage mesh in ocean enclosures.

# Disease

Compared to inshore aquaculture, disease problems currently appear to be much reduced when farming offshore. For example, parasitic infections that occur in mussels cultured offshore are much smaller than those cultured inshore. However, new species are now being farmed offshore although little is known about their ecology and epidemiology. The implications of transmitting pathogens between such farmed species and wild species "remains a large and unanswered question".

Spreading of pathogens between fish stocks is a major issue in disease control. Static offshore cages may help minimize direct spreading, as there may be greater distances between aquaculture production areas. However, development of roaming cage technology could bring about new issues with disease transfer and spread. The high level of carnivorous aquaculture production results in an increased demand for live aquatic animals for production and breeding purposes such as bait, broodstock and milt. This can result in spread of disease across species barriers.

# Employment

Aquaculture is encouraged by many governments as a way to generate jobs and income, particularly when wild fisheries have been run down. However, this may not apply to offshore aquaculture. Offshore aquaculture entails high equipment and supply costs, and therefore will be under severe pressure to lower labor costs through automated production technologies. Employment is likely to expand more at processing facilities than grow-out industries as offshore aquaculture develops.

## Prospects

Norway and the United States are currently (2008) making the main investments in the design of offshore cages.

## FAO

In 2010, the Food and Agriculture Organization (FAO) sub-committee on aquaculture made the following assessments:

"Most Members thought it inevitable that aquaculture will move further offshore if the world is to meet its growing demand for seafood and urged the development of appropriate technologies for its expansion and assistance to developing countries in accessing them [...] Some Members noted that aquaculture may also develop offshore in large inland water bodies and discussion should extend to inland waters as well [...] Some Members suggested caution regarding potential negative impacts when developing offshore aquaculture.

The sub-committee recommended the FAO "should work towards clarifying the technical and legal terminology related to offshore aquaculture in order to avoid confusion."

## Europe

In 2002, the European Commission issued the following policy statement on aquaculture:

"Fish cages should be moved further from the coast, and more research and development of offshore cage technology must be promoted to this end. Experience from outside the aquaculture sector, e.g. with oil platforms, may well feed into the aquaculture equipment sector, allowing for savings in the development costs of technologies."

By 2008, European offshore systems were operating in Norway, Ireland, Italy, Spain, Greece, Cyprus, Malta, Croatia, Portugal and Libya.

In Ireland, as part of their National Development Plan, it is envisioned that over the period 2007–2013, technology associated with offshore aquaculture systems will be developed, including: "sensor systems for feeding, biomass and health monitoring, feed control, telemetry and communications [and] cage design, materials, structural testing and modelling."

## United States

Moving aquaculture offshore into the exclusive economic zone (EEZ) can cause complications with regulations. In the United States, regulatory control of the coastal states generally extends to 3 nm, while federal waters (or EEZ) extend to 200 nm offshore. Therefore, offshore aquaculture can be sited outside the reach of state law but within federal jurisdiction. As of 2010, "all commercial aquaculture facilities have been sited in nearshore waters under state or territorial jurisdiction." However, "unclear regulatory processes" and "technical uncertainties related to working in offshore areas" have hindered progress. The five offshore research projects and commercial operations in the US – in New Hampshire, Puerto Rico, Hawaii and California – are all in federal waters. In June 2011, the *National Sustainable Offshore Aquaculture Act of 2011* was introduced to the House of Representatives "to establish a regulatory system and research program for sustainable offshore aquaculture in the United States exclusive economic zone".

## Current Species

By 2005, offshore aquaculture was present in 25 countries, both as experimental and commercial farms. Market demand means that the most offshore farming efforts are directed towards raising finfish. Two commercial operations in the US, and a third in the Bahamas are using submersible cages to raise high-value carnivorous finfish, such as moi, cobia, and mutton snapper. Submersible cages are also being used in experimental systems for halibut, haddock, cod, and summer flounder in New Hampshire waters, and for amberjack, red drum, snapper, pompano, and cobia in the Gulf of Mexico.

The offshore aquaculture of shellfish grown in suspended culture systems, like scallops and mussels, is gaining ground. Suspended culture systems include methods where the shellfish are grown on a tethered rope or suspended from a floating raft in net containers. Mussels in particular can survive the high physical stress levels which occur in the volatile environments that occur in offshore waters. Finfish species must be feed regularly, but shellfish do not, which can reduce costs. The University of New Hampshire in the US has conducted research on the farming of blue mussels submerged in an open ocean environment. They have found that when farmed in less polluted waters offshore, the mussels develop more flesh with lighter shells.

## Global Status

| Location | Species | Status | Comment |
|---|---|---|---|
| | **Global status of offshore aquaculture** *Aquaculture Collaborative Research Support Program* | | |
| Australia | tuna | C | 10,000 tonnes/year worth A$250 million |
| California | striped bass, California yellowtail, Pacific halibut, abalone | E/C | Attempts to produce from an oil platform |
| Canada | cod, sablefish, mussels, salmon | | Mussels established in eastern Canada |
| Canary Islands | seabass, seabream | | Two cages installed but not now used |
| China | unknown finfish, scallops | E | Small scale experiments on finfish |
| Croatia | tuna | C | 8 offshore cages (1998) |
| Cyprus | seabass, seabream | C | 8 offshore cages (1998) |
| Faeroe Island | | | Failed trials |
| France | seabass, seabream | C | 13 offshore cages (1998) |
| Germany | seaweed, mussels | E | Trials using wind-farms |
| Greece | seabass, seabream | C | |
| Hawaii | amberjack, Pacific threadfin | C | |
| Ireland | Atlantic salmon | E | Various experimental projects |

| Italy | seabass, seabream, tuna | C | |
|---|---|---|---|
| Japan | tuna, mussels | C | Commercial tuna ranching, offshore mussel long-lines. |
| Korea | scallop | | |
| Malta | seabass, seabream, tuna | C | 3 offshore cages (1998) |
| Mexico | tuna | C | |
| Morocco | tuna | C | |
| New Hampshire | Atlantic halibut, cod, haddock, mussels, sea scallops, summer flounder | E/C | Experimental work from the University of New Hampshire, two commercial mussel sites |
| New Zealand | mussels | | About to become operational |
| Panama | tuna | C | |
| Puerto Rico | cobia, snapper | C | |
| Spain | seabass, seabream | C | Government assisting trials |
| Turkey | seabass, seabream | C | |
| Washington | sablefish | C | |
| Taiwan | cobia | C | 3,000 tonnes (2001) |

Status: E = Experimental, C = Commercial

# Oyster Farming

Harvesting oysters from the pier at Cancale, Brittany, France 2005

Oyster farming is an aquaculture (or mariculture) practice in which oysters are raised for human consumption. Oyster farming was practiced by the ancient Romans as early as the 1st century BC on the Italian peninsula and later in Britain for export to Rome. The French oyster industry has relied on aquacultured oysters since the late 18th century.

## History

Oyster farming was practiced by the ancient Romans as early as the 1st century BC on the Italian peninsula. With the Barbarian invasions the oyster farming in the Mediterranean and the Atlantic came to an end.

Oyster harvesting using rakes (top) and sail driven dredges (bottom). From *L'Encyclpédie* of 1771

Oyster farming boats in Morbihan, France

Harvesting oysters from beds by hand in Willapa Bay, United States

Oysters farmed in baskets on Prince Edward Island, Canada

Boats used for culturing oysters (circa 1920) in the Gironde estuary, France

Flat bottomed oyster-boat with oyster-bags in Chaillevette, France

In 1852 Monsieur de Bon started to re-seed the oyster beds by collecting the oyster spawn using makeshift catchers. An important step to the modern oyster farming was the oyster farm built by Hyacinthe Boeuf in the Ile de Ré. After obtaining the rights to a part of the coast he built a wall to make a reservoir and to break the strength of the current. Some time later the wall was covered with spat coming spontaneously from the sea which gave 2000 baby oysters per square metre.

## Varieties of Farmed Oysters

Commonly farmed food oysters include the Eastern oyster *Crassostrea virginica*, the Pacific oyster *Crassostrea gigas*, Belon oyster *Ostrea edulis*, the Sydney rock oyster *Saccostrea glomerata*, and the Southern mud oyster *Ostrea angasi*.

## Cultivation

Oysters naturally grow in estuarine bodies of brackish water. When farmed, the temperature and salinity of the water are controlled (or at least monitored), so as to induce spawning and fertilization, as well as to speed the rate of maturation – which can take several years.

Three methods of cultivation are commonly used. In each case oysters are cultivated to the size of "spat," the point at which they attach themselves to a substrate. The substrate is known as a "cultch" (also spelled "cutch" or "culch"). The loose spat may be allowed to mature further to form "seed" oysters with small shells. In either case (spat or seed stage), they are then set out to mature. The maturation technique is where the cultivation method choice is made.

In one method the spat or seed oysters are distributed over existing oyster beds and left to mature naturally. Such oysters will then be collected using the methods for fishing wild oysters, such as dredging.

In the second method the spat or seed may be put in racks, bags, or cages (or they may be glued in threes to vertical ropes) which are held above the bottom. Oysters cultivated in this manner may be harvested by lifting the bags or racks to the surface and removing mature oysters, or simply retrieving the larger oysters when the enclosure is exposed at low tide. The latter method may avoid losses to some predators, but is more expensive.

In the third method the spat or seed are placed in a cultch within an artificial maturation tank. The maturation tank may be fed with water that has been especially prepared for the purpose of accelerating the growth rate of the oysters. In particular the temperature and salinity of the water may be altered somewhat from nearby ocean water. The carbonate minerals calcite and aragonite in the water may help oysters develop their shells faster and may also be included in the water processing prior to introduction to the tanks. This latter cultivation technique may be the least susceptible to predators and poaching, but is the most expensive to build and to operate. The Pacific oyster *C. gigas* is the species most commonly used with this type of farming.

Oyster culture using tiles as cultch. Taken from The Illustrated London News 1881

Purpose made oyster baskets

Working on oysters at Belon, Brittany, France 2005

Oyster farm in South Australia

Oyster shucking at Lau Fau Shan, Hong Kong

Here in Yerseke, Netherlands, oysters are kept in large oyster pits after "harvesting", until they are sold. Seawater is pumped in and out, simulating the tide

## Boats

During the nineteenth century in the United States, various shallow draft sailboat designs were developed for oystering in Chesapeake Bay. These included the bugeye, log canoe, pungy, sharpie and skipjack. During the 1880s, a powerboat called the Chesapeake Bay deadrise was also developed.

Since 1977, several boat builders in Brittany have built specialized amphibious vehicles for use in the area's mussel and oyster farming industries. The boats are made of aluminium, are relatively flat-bottomed, and have three, four, or six wheels, depending on the size of the boat. When the tide is out the boats can run on the tidal flats using their wheels. When the tide is in, they use a propeller to move themselves through the water. Oyster farmers in Jersey make use of similar boats. Currently, *Constructions Maritimes du Vivier Amphibie* has a range of models.

## Environmental Impact

The farming of oysters and other shellfish is relatively benign or even restorative environmentally, and holds promise for relieving pressure on land-based protein sources. Restoration of oyster populations are encouraged for the ecosystem services they provide, including water quality maintenance, shoreline protection and sediment stabilization, nutrient cycling and sequestration, and habitat for other organisms. A native Olympia oyster restoration project is taking place in Liberty Bay, Washington, and numerous oyster restoration projects are underway in the Chesapeake Bay. In the U.S., Delaware is the only East Coast state without oyster aquaculture, but making aquaculture a state-controlled industry of leasing water by the acre for commercial harvesting of shellfish is being considered. Supporters of Delaware's legislation to allow aquaculture cite revenue, job creation, and nutrient cycling benefits. It is estimated that one acre can produce nearly 750,000 oysters, which could filter between 15 and 40 million gallons of water daily.

Other sources state that a single oyster can filter 24–96 liters a day(1–4 liters per hour). With 750,000 oysters in one acre, 18,000,000-72,000,000 liters of water can be filtered, removing most forms of particulate matter suspended in the water column. The particulate matter oysters remove are sand, clay, silt, detritus, and phytoplankton. These particulates all could possibly contain harmful contamination that originates from anthropogenic sources (the land or directly flowing into the body of water). Instead of becoming ingested by other filter feeders that are then digested by bigger organisms, oysters can sequester these possibly harmful pollutants, and excrete them into the sediment at the bottom of waterways. To remove these contaminants from the sediment, species of seaweed can be added to take up these contaminants in their plant tissues that could be removed and taken to a contained area where the contamination is benign to the surrounding environment.

## Predators, Diseases and Pests

Oyster predators include starfish, oyster drill snails, stingrays, Florida stone crabs, birds, such as oystercatchers and gulls, and humans.

Pathogens that can affect either farmed *C. virginica* or *C. gigas* oysters include *Perkinsus marinus* (Dermo) and *Haplosporidium nelsoni* (MSX). However, *C. virginica* are much more susceptible to Dermo or MSX infections than are the *C. gigas* species of oyster. Pathogens of *O. edulis* oysters include *Marteilia refringens* and *Bonamia ostreae*. In the north Atlantic Ocean, oyster crabs may live in an endosymbiotic commensal relationship within a host oyster. Since oyster crabs are considered a food delicacy they may not be removed from young farmed oysters, as they can themselves be harvested for sale.

Polydorid polychaetes are known as pests of cultured oysters.

# Integrated Multi-trophic Aquaculture

Integrated multi-trophic aquaculture (IMTA) provides the byproducts, including waste, from one aquatic species as inputs (fertilizers, food) for another. Farmers combine fed aquaculture (e.g., fish, shrimp) with inorganic extractive (e.g., seaweed) and organic extractive (e.g., shellfish) aquaculture to create balanced systems for environment remediation (biomitigation), economic stability (improved output, lower cost, product diversification and risk reduction) and social acceptability (better management practices).

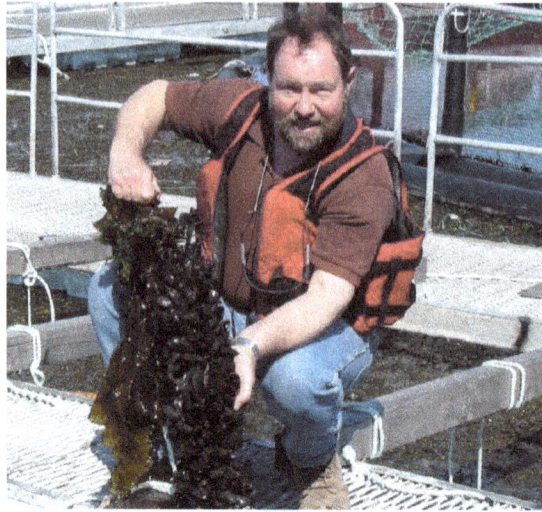

Blue mussels (*Mytilus edulis*) cultivated in proximity to Atlantic salmon (*Salmo salar*) in the Bay of Fundy, Canada. *Note the salmon cage (polar circle) in the background.*

Selecting appropriate species and sizing the various populations to provide necessary ecosystem functions allows the biological and chemical processes involved to achieve a stable balance, mutually benefiting the organisms and improving ecosystem health.

Ideally, the co-cultured species each yield valuable commercial "crops". IMTA can synergistically increase total output, even if some of the crops yield less than they would, short-term, in a monoculture.

## Terminology and Related Approaches

"Integrated" refers to intensive and synergistic cultivation, using water-borne nutrient and energy transfer. "Multi-trophic" means that the various species occupy different trophic levels, i.e., different (but adjacent) links in the food chain.

IMTA is a specialized form of the age-old practice of aquatic polyculture, which was the co-culture of various species, often without regard to trophic level. In this broader case, the organisms may share biological and chemical processes that are minimally complementary, potentially leading to significant ecosystem shifts/damage. Some traditional systems did culture species that occupied multiple niches within the same pond, but with limited intensity and management.

The more general term "Integrated Aquaculture" is used to describe the integration of monocultures through water transfer. The terms "IMTA" and "integrated aquaculture" differ primarily in their precision and are sometimes interchanged. Aquaponics, fractionated aquaculture, integrated agriculture-aquaculture systems, integrated peri-urban-aquaculture systems, and integrated fisheries-aquaculture systems are variations on the IMTA concept.

## Range of Approaches

Today, low-intensity traditional/incidental multi-trophic aquaculture is much more common than modern IMTA. Most are relatively simple, such as fish/seaweed/shellfish.

True IMTA can be land-based, using ponds or tanks, or even open-water marine or freshwater systems. Implementations have included species combinations such as shellfish/shrimp, fish/seaweed/shellfish, fish/seaweed, fish/shrimp and seaweed/shrimp.

IMTA in open water (offshore cultivation) can be done by the use of buoys with lines on which the seaweed grows. The buoys/lines are placed next to the fishnets or cages in which the fish grows. This method is already used commercially in Norway, Scotland, and Ireland.

In the future, systems with other components for additional functions, or similar functions but different size brackets of particles, are likely. Multiple regulatory issues remain open.

## Modern History of Land-based Systems

Ryther and co-workers created modern, integrated, intensive, land mariculture. They originated, both theoretically and experimentally, the integrated use of extractive organisms—shellfish, microalgae and seaweeds—in the treatment of household effluents, descriptively and with quantitative results. A domestic wastewater effluent, mixed with seawater, was the nutrient source for phytoplankton, which in turn became food for oysters and clams. They cultivated other organisms in a food chain rooted in the farm's organic sludge. Dissolved nutrients in the final effluent were filtered by seaweed (mainly Gracilaria and Ulva) biofilters. The value of the original organisms grown on human waste effluents was minimal.

In 1976, Huguenin proposed adaptations to the treatment of intensive aquaculture effluents in both inland and coastal areas. Tenore followed by integrating with their system of carnivorous fish and the macroalgivore abalone.

In 1977, Hughes-Games described the first practical marine fish/shellfish/phytoplankton culture, followed by Gordin, et al., in 1981. By 1989, a semi-intensive (1 kg fish/m$^{-3}$) seabream and grey mullet pond system by the Gulf of Aqaba (Eilat) on the Red Sea supported dense diatom populations, excellent for feeding oysters. Hundreds of kilos of fish and oysters cultured here were sold. Researchers also quantified the water quality parameters and nutrient budgets in (5 kg fish m$^{-3}$) green water seabream ponds. The phytoplankton generally maintained reasonable water quality and converted on average over half the waste nitrogen into algal biomass. Experiments with intensive bivalve cultures yielded high bivalve growth rates. This technology supported a small farm in southern Israel.

## Sustainability

IMTA promotes economic and environmental sustainability by converting byproducts and uneaten feed from fed organisms into harvestable crops, thereby reducing eutrophication, and increasing economic diversification.

Properly managed multi-trophic aquaculture accelerates growth without detrimental side-effects. This increases the site's ability to assimilate the cultivated organisms, thereby reducing negative environmental impacts.

IMTA enables farmers to diversify their output by replacing purchased inputs with byproducts from lower trophic levels, often without new sites. Initial economic research suggests that IMTA can increase profits and can reduce financial risks due to weather, disease and market fluctuations. Over a dozen studies have investigated the economics of IMTA systems since 1985.

## Nutrient Flow

Typically, carnivorous fish or shrimp occupy IMTA's higher trophic levels. They excrete soluble ammonia and phosphorus (orthophosphate). Seaweeds and similar species can extract these inorganic nutrients directly from their environment. Fish and shrimp also release organic nutrients which feed shellfish and deposit feeders.

Species such as shellfish that occupy intermediate trophic levels often play a dual role, both filtering organic bottom-level organisms from the water and generating some ammonia. Waste feed may also provide additional nutrients; either by direct consumption or via decomposition into individual nutrients. In some projects, the waste nutrients are also gathered and reused in the food given to the fish in cultivation. This can happen by processing the seaweed grown into food.

## Recovery Efficiency

Nutrient recovery efficiency is a function of technology, harvest schedule, management, spatial configuration, production, species selection, trophic level biomass ratios, natural food availability, particle size, digestibility, season, light, temperature, and water flow. Since these factors significantly vary by site and region, recovery efficiency also varies.

In a hypothetical family-scale fish/microalga /bivalve/seaweed farm, based on pilot scale data, at least 60% of nutrient input reached commercial products, nearly three times more than in modern net pen farms. Expected average annual yields of the system for a hypothetical 1 hectare (2.5 acres) were 35 tonnes (34 long tons; 39 short tons) of seabream, 100 tonnes (98 long tons; 110 short tons) of bivalves and 125 tonnes (123 long tons; 138 short tons) of seaweeds. These results required precise water quality control and attention to suitability for bivalve nutrition, due to the difficulty in maintaining consistent phytoplanton populations.

Seaweeds' nitrogen uptake efficiency ranges from 2-100% in land-based systems. Uptake efficiency in open-water IMTA is unknown.

## Food Safety and Quality

Feeding the wastes of one species to another has the potential for contamination, although this has yet to be observed in IMTA systems. Mussels and kelp growing adjacent to Atlantic salmon cages in the Bay of Fundy have been monitored since 2001 for contamination by medicines, heavy metals, arsenic, PCBs and pesticides. Concentrations are consistently either non-detectable or well below regulatory limits established by the Canadian Food Inspection Agency, the United States Food and Drug Administration and European Community Directives. Taste testers indicate that these mussels are free of "fishy" taste and aroma and could not distinguish them from "wild" mussels. The mussels' meat yield is significantly higher, reflecting the increase in nutrient availability. Recent findings suggest mussels grown adjacent to salmon farms are advantageous for winter harvest because they maintain high meat weight and condition index (meat to shell ratio). This finding is of particular interest because the Bay of Fundy, where this research was conducted, produces low condition index mussels during winter months in monoculture situations, and seasonal presence of Paralytic Shellfish Poisoning (PSP) typically restricts mussel harvest to the winter months.

## Selected Projects

Historic and ongoing research projects include:

## Asia

Japan, China, South Korea, Thailand, Vietnam, Indonesia, Bangladesh, etc. have co-cultured aquatic species for centuries in marine, brackish and fresh water environments. Fish, shellfish and seaweeds have been cultured together in bays, lagoons and ponds. Trial and error has improved integration over time. The proportion of Asian aquaculture production that occurs in IMTA systems is unknown.

After the 2004 tsunami, many of the shrimp farmers in Aceh Province of Indonesia and Ranong Province of Thailand were trained in IMTA. This has been especially important as the mono-culture of marine shrimp was widely recognized as unsustainable. Production of tilapia, mud crabs, seaweeds, milkfish, and mussels have been incorporated. AquaFish Collaborative Research Support Program

## Canada

### Bay of Fundy

Industry, academia and government are collaborating here to expand production to commercial scale. The current system integrates Atlantic salmon, blue mussels and kelp; deposit feeders are under consideration. AquaNet (one of Canada's Networks of Centres of Excellence) funded phase one. The Atlantic Canada Opportunities Agency is funding phase two. The project leaders are Thierry Chopin (University of New Brunswick in Saint John) and Shawn Robinson (Department of Fisheries and Oceans, St. Andrews Biological Station).

### Pacific SEA-lab

Pacific SEA-lab is researching and is licensed for the co-culture of sablefish, scallops, oysters, blue mussels, urchins and kelp. "SEA" stands for Sustainable Ecological Aquaculture. The project aims to balance four species.The project is headed by Stephen Cross under a British Columbia Innovation Award at the University of Victoria Coastal Aquaculture Research & Training (CART) network.

## Chile

The i-mar Research Center at the Universidad de Los Lagos, in Puerto Montt is working to reduce the environmental impact of intensive salmon culture. Initial research involved trout, oysters and seaweeds. Present research is focusing on open waters with salmon, seaweeds and abalone. The project leader is Alejandro Buschmann.

## Israel

### Sea or Marine Enterprises Ltd.

SeaOr Marine Enterprises Ltd., which operated for several years on the Israeli Mediterranean coast, north of Tel Aviv, cultured marine fish (gilthead seabream), seaweeds (Ulva and Gracilaria) and Japanese abalone. Its approach leveraged local climate, and recycled fish waste products into seaweed biomass, which was fed to the abalone. It also effectively purified the water sufficiently to allow the water to be recycled to the fishponds and to meet point-source effluent environmental regulations.

## PGP Ltd.

PGP Ltd. is a small farm in Southern Israel. It cultures marine fish, microalgae, bivalves and Artemia. Effluents from seabream and seabass collect in sedimentation ponds, where dense populations of microalgae—mostly diatoms—develop. Clams, oysters and sometimes Artemia filter the microalgae from the water, producing a clear effluent. The farm sells the fish, bivalves and Artemia.

## The Netherlands

In the Netherlands, Willem Brandenburg of UR Wageningen (Plant Sciences Group) has established the first seaweed farm in the Netherlands. The farm is called "De Wierderij" and is used for research.

## South Africa

Three farms grow seaweeds for feed in abalone effluents in land-based tanks. Up to 50% of re-circulated water passes through the seaweed tanks. Somewhat uniquely, neither fish nor shrimp comprise the upper trophic species. The motivation is to avoid over-harvesting natural seaweed beds and red tides, rather than nutrient abatement. These commercial successes developed from research collaboration between Irvine and Johnson Cape Abalone and scientists from the University of Cape Town and the University of Stockholm.

## United Kingdom

The Scottish Association for Marine Science, in Oban is developing co-cultures of salmon, oysters, sea urchins, and brown and red seaweeds via several projects. Research focuses on biological and physical processes, as well as production economics and implications for coastal zone management. Researchers include: M. Kelly, A. Rodger, L. Cook, S. Dworjanyn, and C. Sanderson.

## Bangladesh

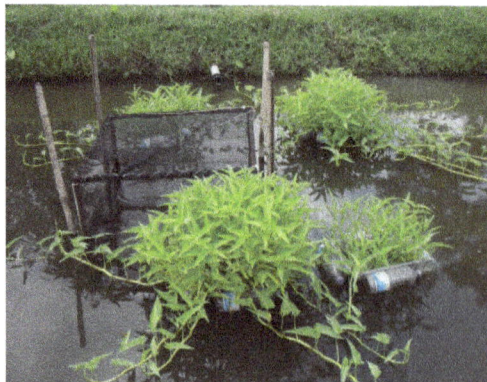

IMTA systems in freshwater pond

Indian carps and stinging catfish are cultured in Bangladesh, but the methods could be more productive. The pond and cage cultures used are based only on the fish. They don't take advantage of the productivity increases that could take place if other trophic levels were included. Expensive artificial feeds are used, partly to supply the fish with protein. These costs could be reduced if freshwater snails, such as *Viviparus bengalensis*, were simultaneously cultured, thus increasing the available protein. The organic and inorganic wastes produced as a byproduct of culturing could also be minimized by integrating freshwater snail and aquatic plants, such as water spinach,respectively.

## Gallery

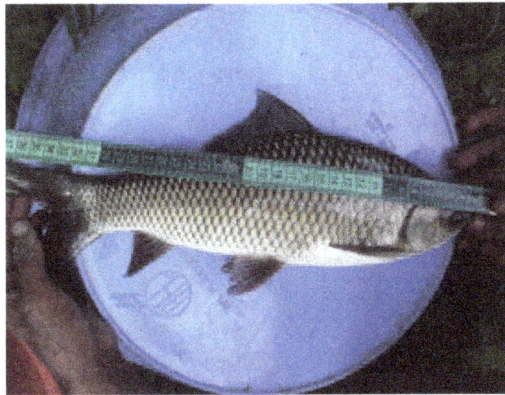

Carp (Labeo rohita) produced in IMTA pond

Off-bottom snail grown on bamboo split in IMTA

Snail produced on pond bottom of IMTA

Collection of water spinach and snail from IMTA pond

Produced shing in cage in IMTA

# References

- Arnason, Ragnar (2001) Ocean Ranching in Japan In: The Economics of Ocean Ranching: Experiences, Outlook and Theory, FAO, Rome. ISBN 92-5-104631-X.

- Black, K. D. (2001). "Mariculture, Environmental, Economic and Social Impacts of". In Steele, John H.; Thorpe, Steve A.; Turekian, Karl K. Encyclopedia of Ocean Sciences. Academic Press. pp. 1578–1584. doi:10.1006/rwos.2001.0487. ISBN 9780122274305.

- Sturrock H, Newton R, Paffrath S, Bostock J, Muir J, Young J, Immink A and Dickson M (2008) Part 2: Characterisation of emerging aquaculture systems In: Prospective Analysis of the Aquaculture Sector in the EU, European Commission, EUR 23409 EN/2. ISBN 978-92-79-09442-2.

- Martinez-Cordero FJ (2007) "Socioeconomic Aspects of Species and Systems Selection for Sustainable Aquaculture" pp. 225–239. In: Leung P, Lee C and O'Bryen P (Eds.) Species and system selection for sustainable aquaculture, John Wiley & Sons. ISBN 978-0-8138-2691-2. doi:10.1002/9780470277867.

- Bostock J, Muir J, Young J, Newton R and Paffrath S (2008) Part 1: Synthesis report In: Prospective Analysis of the Aquaculture Sector in the EU, European Commission, EUR 23409 EN/1. ISBN 978-92-79-09441-5.

- Higginbotham, James Arnold (1997-01-01). Piscinae: Artificial Fishponds in Roman Italy. UNC Press Books. ISBN 9780807823293.

- "Information Memorandum, 2013 Ranching of Greenlip Abalone, Flinders Bay – Western Australia" (PDF). Ocean Grown Abalone. Ocean Grown Abalone. Retrieved 23 April 2016.

- Fitzgerald, Bridget (28 August 2014). "First wild abalone farm in Australia built on artificial reef". Australian Broadcasting Corporation Rural. Australian Broadcasting Corporation. Retrieved 23 April 2016.

- Brown, Ashton (June 10, 2013). "'Aquaculture' shellfish harvesting bill moves forward". Delaware State News. Retrieved June 11, 2013.

- Borgatti, Rachel; Buck, Eugene H. (December 13, 2004). "Open Ocean Aquaculture" (PDF). Congressional Research Service. Retrieved April 10, 2010.

- McAvoy, Audrey (October 24, 2009). "Hawaii regulators approve first US tuna farm". The Associated Press. Retrieved April 9, 2010.

**3**

# Classification of Aquaculture According to Species

Aquaculture has been particularly effective in the cultivation and harvest of certain marine species. Some of the more popular species that have been brought into the fold of aquaculture cultivation are seaweed, catfish, algae, giant kelp, sea cucumbers etc. The topics discussed in the chapter are of great importance to broaden the existing knowledge on aquaculture.

## Algaculture

Algaculture is a form of aquaculture involving the farming of species of algae.

The majority of algae that are intentionally cultivated fall into the category of microalgae (also referred to as phytoplankton, microphytes, or planktonic algae). Macroalgae, commonly known as seaweed, also have many commercial and industrial uses, but due to their size and the specific requirements of the environment in which they need to grow, they do not lend themselves as readily to cultivation (this may change, however, with the advent of newer seaweed cultivators, which are basically algae scrubbers using upflowing air bubbles in small containers).

Commercial and industrial algae cultivation has numerous uses, including production of food ingredients such as omega-3 fatty acids or natural food colorants and dyes, food, fertilizer, bioplastics, chemical feedstock (raw material), pharmaceuticals, and algal fuel, and can also be used as a means of pollution control.

### Growing, Harvesting, and Processing Algae

### Monoculture

Most growers prefer monocultural production and go to considerable lengths to maintain the purity of their cultures. With mixed cultures, one species comes to dominate over time and if a non-dominant species is believed to have particular value, it is necessary to obtain pure cultures in order to cultivate this species. Individual species cultures are also needed for research purposes.

A common method of obtaining pure cultures is serial dilution. Cultivators dilute either a wild sample or a lab sample containing the desired algae with filtered water and introduce small aliquots (measures of this solution) into a large number of small growing containers. Dilution follows a microscopic examination of the source culture that predicts that a few of the growing containers contain a single cell of the desired species. Following a suitable period on a light table, cultivators again use the microscope to identify containers to start larger cultures.

Another approach is to use a special medium which excludes other organisms, including invasive algae. For example, *Dunaliella* is a commonly grown genus of microalgae which flourishes in extremely salty water that few other organisms can tolerate.

Alternatively, mixed algae cultures can work well for larval mollusks. First, the cultivator filters the sea water to remove algae which are too large for the larvae to eat. Next, the cultivator adds nutrients and possibly aerates the result. After one or two days in a greenhouse or outdoors, the resulting thin soup of mixed algae is ready for the larvae. An advantage of this method is low maintenance.

## Growing Algae

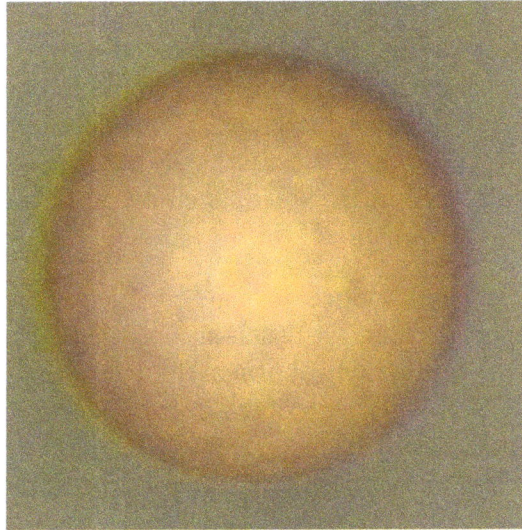

Microalgae is used to culture brine shrimp, which produce dormant eggs (pictured). The eggs can then be hatched on demand and fed to cultured fish larvae and crustaceans.

Water, carbon dioxide, minerals and light are all important factors in cultivation, and different algae have different requirements. The basic reaction for algae growth in water is carbon dioxide + light energy + water = glucose + oxygen + water. This is called *autotrophic* growth. It is also possible to grow certain types of algae without light, these types of algae consume sugars (such as glucose). This is called *heterotrophic* growth.

## Temperature

The water must be in a temperature range that will support the specific algal species being grown mostly between 25 to 35 degrees C.

## Light and Mixing

In a typical algal-cultivation system, such as an open pond, light only penetrates the top 3 to 4 inches (76–102 mm) of the water, though this depends on the algae density. As the algae grow and multiply, the culture becomes so dense that it blocks light from reaching deeper into the water. Direct sunlight is too strong for most algae, which can use only about $\frac{1}{10}$ the amount of light they receive from direct sunlight; however, exposing an algae culture to direct sunlight (rather than shading it) is often the best course for strong growth, as the algae underneath the surface get more light.

To use deeper ponds, growers agitate the water, circulating the algae so that it does not remain on the surface. Paddle wheels can stir the water and compressed air coming from the bottom lifts algae from the lower regions. Agitation also helps prevent over-exposure to the sun.

Another means of supplying light is to place the light *in* the system. Glow plates made from sheets of plastic or glass and placed within the tank offer precise control over light intensity, and distribute it more evenly. They are seldom used, however, due to high cost.

## Odor and Oxygen

The odor associated with bogs, swamps, indeed any stagnant waters, can be due to oxygen depletion caused by the decay of deceased algal blooms. Under anoxic conditions, the bacteria inhabiting algae cultures break down the organic material and produce hydrogen sulfide and ammonia which causes the odor. This hypoxia often results in the death of aquatic animals. In a system where algae is intentionally cultivated, maintained, and harvested, neither eutrophication nor hypoxia are likely to occur.

Some living algae and bacteria, also produce odorous chemicals, particularly certain (cyanobacteria) (previously classed as blue-green algae) such as *Anabaena*. The most well-known of these odor-causing chemicals are MIB (2-methylisoborneol) and geosmin. They give a musty or earthy odor that can be quite strong. Eventual death of the cyanobacteria releases additional gas that is trapped in the cells. These chemicals are detectable at very low levels, in the parts per billion range, and are responsible for many "taste and odor" issues in drinking water treatment and distribution. Cyanobacteria can also produce chemical toxins that have been a problem in drinking water.

## Nutrients

Nutrients such as nitrogen (N), phosphorus (P), and potassium (K) serve as fertilizer for algae, and are generally necessary for growth. Silica and iron, as well as several trace elements, may also be considered important marine nutrients as the lack of one can limit the growth of, or productivity in, a given area. Carbon dioxide is also essential; usually an input of $CO_2$ is required for fast-paced algal growth. These elements must be dissolved into the water, in bio-available forms, for algae to grow.

## Pond and Bioreactor Cultivation Methods

Algae can be cultured in open ponds (such as raceway-type ponds and lakes) and photobioreactors. Raceway ponds may be less expensive.

## Open Ponds

Raceway-type ponds and lakes are open to the elements. Open ponds are highly vulnerable to contamination by other microorganisms, such as other algal species or bacteria. Thus cultivators usually choose closed systems for monocultures. Open systems also do not offer control over temperature and lighting. The growing season is largely dependent on location and, aside from tropical areas, is limited to the warmer months.

Raceway pond used to cultivate microalgae. The water is kept in constant motion with a powered paddle wheel.

Open pond systems are cheaper to construct, at the minimum requiring only a trench or pond. Large ponds have the largest production capacities relative to other systems of comparable cost. Also, open pond cultivation can exploit unusual conditions that suit only specific algae. For instance, *Dunaliella salina* grow in extremely salty water; these unusual media exclude other types of organisms, allowing the growth of pure cultures in open ponds. Open culture can also work if there is a system of harvesting only the desired algae, or if the ponds are frequently re-inoculated before invasive organisms can multiply significantly. The latter approach is frequently employed by Chlorella farmers, as the growth conditions for Chlorella do not exclude competing algae.

The former approach can be employed in the case of some chain diatoms since they can be filtered from a stream of water flowing through an outflow pipe. A "pillow case" of a fine mesh cloth is tied over the outflow pipe allowing other algae to escape. The chain diatoms are held in the bag and feed shrimp larvae (in Eastern hatcheries) and inoculate new tanks or ponds.

Enclosing a pond with a transparent or translucent barrier effectively turns it into a greenhouse. This solves many of the problems associated with an open system. It allows more species to be grown, it allows the species that are being grown to stay dominant, and it extends the growing season – if heated, the pond can produce year round. Open race way ponds were used for removal of lead using live *Spirulina (Arthospira) sp.*

## Photobioreactors

Algae can also be grown in a photobioreactor (PBR). A PBR is a bioreactor which incorporates a light source. Virtually any translucent container could be called a PBR; however, the term is more commonly used to define a closed system, as opposed to an open tank or pond.

Because PBR systems are closed, the cultivator must provide all nutrients, including $CO_2$.

A PBR can operate in "batch mode", which involves restocking the reactor after each harvest, but it is also possible to grow and harvest continuously. Continuous operation requires precise control of all elements to prevent immediate collapse. The grower provides sterilized water, nutrients, air, and carbon dioxide at the correct rates. This allows the reactor to operate for long periods. An advantage is that algae that grows in the "log phase" is generally of higher nutrient content than old "senescent" algae. Algal culture is the culturing of algae in ponds or other resources. Maximum productivity occurs when the "exchange rate" (time to exchange one volume of liquid) is equal to the "doubling time" (in mass or volume) of the algae.

Different types of PBRs include:

- Tanks

- Polyethylene sleeves or bags

- Glass or plastic tubes.

## Harvesting

A seaweed farmer in Nusa Lembongan gathers edible seaweed that has grown on a rope.

Algae can be harvested using microscreens, by centrifugation, by flocculation and by froth flotation.

Interrupting the carbon dioxide supply can cause algae to flocculate on its own, which is called "autoflocculation".

"Chitosan", a commercial flocculant, more commonly used for water purification, is far more expensive. The powdered shells of crustaceans are processed to acquire chitin, a polysaccharide found in the shells, from which chitosan is derived via de-acetylation. Water that is more brackish, or saline requires larger amounts of flocculant. Flocculation is often too expensive for large operations.

Alum and ferric chloride are other chemical flocculants.

In froth flotation, the cultivator aerates the water into a froth, and then skims the algae from the top.

Ultrasound and other harvesting methods are currently under development.

## Oil Extraction

Algae oils have a variety of commercial and industrial uses, and are extracted through a variety of methods. Estimates of the cost to extract oil from microalgae vary, but are likely to be around three times higher than that of extracting palm oil.

## Physical Extraction

In the first step of extraction, the oil must be separated from the rest of the algae. The simplest method is mechanical crushing. When algae is dried it retains its oil content, which then can be "pressed" out with an oil press. Different strains of algae warrant different methods of oil pressing, including the use of screw, expeller and piston. Many commercial manufacturers of vegetable oil use a combination of mechanical pressing and chemical solvents in extracting oil. This use is often also adopted for algal oil extraction.

Osmotic shock is a sudden reduction in osmotic pressure, this can cause cells in a solution to rupture. Osmotic shock is sometimes used to release cellular components, such as oil.

Ultrasonic extraction, a branch of sonochemistry, can greatly accelerate extraction processes. Using an ultrasonic reactor, ultrasonic waves are used to create cavitation bubbles in a solvent material. When these bubbles collapse near the cell walls, the resulting shock waves and liquid jets cause those cells walls to break and release their contents into a solvent. Ultrasonication can enhance basic enzymatic extraction. The combination "sonoenzymatic treatment" accelerates extraction and increases yields.

## Chemical Extraction

Chemical solvents are often used in the extraction of the oils. The downside to using solvents for oil extraction are the dangers involved in working with the chemicals. Care must be taken to avoid exposure to vapors and skin contact, either of which can cause serious health damage. Chemical solvents also present an explosion hazard.

A common choice of chemical solvent is hexane, which is widely used in the food industry and is relatively inexpensive. Benzene and ether can also separate oil. Benzene is classified as a carcinogen.

Another method of chemical solvent extraction is Soxhlet extraction. In this method, oils from the algae are extracted through repeated washing, or percolation, with an organic solvent such as hexane or petroleum ether, under reflux in a special glassware. The value of this technique is that the solvent is reused for each cycle.

Enzymatic extraction uses enzymes to degrade the cell walls with water acting as the solvent. This makes fractionation of the oil much easier. The costs of this extraction process are estimated to be much greater than hexane extraction. The enzymatic extraction can be supported by ultrasonication. The combination "sonoenzymatic treatment" causes faster extraction and higher oil yields.

Supercritical $CO_2$ can also be used as a solvent. In this method, $CO_2$ is liquefied under pressure and heated to the point that it becomes supercritical (having properties of both a liquid and a gas), allowing it to act as a solvent.

Other methods are still being developed, including ones to extract specific types of oils, such as those with a high production of long-chain highly unsaturated fatty acids.

## Algal Culture Collections

Specific algal strains can be acquired from algal culture collections, with over 500 culture collections registered with the World Federation for Culture Collections.

## Uses of Algae

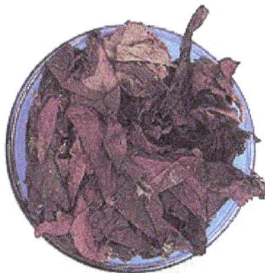

Dulse is one of many edible algae.

## Food

Several species of algae are raised for food.

- Purple laver (*Porphyra*) is perhaps the most widely domesticated marine algae. In Asia it is used in nori (Japan) and gim (Korea). In Wales, it is used in laverbread, a traditional food, and in Ireland it is collected and made into a jelly by stewing or boiling. Preparation also can involve frying or heating the fronds with a little water and beating with a fork to produce a pinkish jelly. Harvesting also occurs along the west coast of North America, and in Hawaii and New Zealand.

- Dulse (*Palmaria palmata*) is a red species sold in Ireland and Atlantic Canada. It is eaten raw, fresh, dried, or cooked like spinach.

- Spirulina (*Arthrospira platensis*) is a blue-green microalgae with a long history as a food source in East Africa and pre-colonial Mexico. Spirulina is high in protein and other nutrients, finding use as a food supplement and for malnutrition. Spirulina thrives in open systems and commercial growers have found it well-suited to cultivation. One of the largest production sites is Lake Texcoco in central Mexico. The plants produce a variety of nutrients and high amounts of protein. Spirulina is often used commercially as a nutritional supplement.

- Chlorella, another popular microalgae, has similar nutrition to spirulina. Chlorella is very popular in Japan. It is also used as a nutritional supplement with possible effects on metabolic rate. Some allege that Chlorella can reduce mercury levels in humans (supposedly by chelation of the mercury to the cell wall of the organism).

- Irish moss (*Chondrus crispus*), often confused with *Mastocarpus stellatus*, is the source of carrageenan, which is used as a stiffening agent in instant puddings, sauces, and dairy products such as ice cream. Irish moss is also used by beer brewers as a fining agent.

- Sea lettuce (*Ulva lactuca*), is used in Scotland where it is added to soups and salads. Dabberlocks or badderlocks (*Alaria esculenta*) is eaten either fresh or cooked in Greenland, Iceland, Scotland and Ireland.

- Aphanizomenon flos-aquae is a cyanobacteria similar to spirulina, which is used as a nutritional supplement.

- Extracts and oils from algae are also used as additives in various food products. The plants also produce Omega-3 and Omega-6 fatty acids, which are commonly found in fish oils, and which have been shown to have positive health benefits.

- Sargassum species are an important group of seaweeds. These algae have many phlorotannins.

- Cochayuyo (Durvillaea Antarctica) is eaten in salads and ceviche in Peru and Chile.

## Fertilizer and Agar

For centuries seaweed has been used as fertilizer. It is also an excellent source of potassium for manufacture of potash and potassium nitrate.

Both microalgae and macroalgae are used to make agar.

## Pollution Control

With concern over global warming, new methods for the thorough and efficient capture of $CO_2$ are being sought out. The carbon dioxide that a carbon-fuel burning plant produces can feed into open or closed algae systems, fixing the $CO_2$ and accelerating algae growth. Untreated sewage can supply additional nutrients, thus turning two pollutants into valuable commodities.

Algae cultivation is under study for uranium/plutonium sequestration and purifying fertilizer run-off.

## Energy Production

Business, academia and governments are exploring the possibility of using algae to make gasoline, diesel and other fuels. Algae itself may be used as a biofuel, and additionally be used to create hydrogen.

## Other Uses

Chlorella, particularly a transgenic strain which carries an extra mercury reductase gene, has been studied as an agent for environmental remediation due to its ability to reduce Hg2+ to the less toxic elemental mercury.

Cultivated algae serve many other purposes, including cosmetics, animal feed, bioplastic production, dyes and colorant production, chemical feedstock production, and pharmaceutical ingredients.

# Seaweed Farming

A seaweed farmer in Nusa Lembongan gathers edible seaweed that has grown on a rope.

Seaweed farming is the practice of cultivating and harvesting seaweed. In its simplest form, it consists of the management of naturally found batches. In its most advanced form, it consists of fully controlling the life cycle of the algae. The main food species grown by aquaculture in

Japan, China and Korea include *Gelidium*, *Pterocladia*, *Porphyra*, and *Laminaria*. Seaweed farming has frequently been developed as an alternative to improve economic conditions and to reduce fishing pressure and over exploited fisheries. Seaweeds have been harvested throughout the world as a food source as well as an export commodity for production of agar and carrageenan products.

## History

Seaweed farming began in Japan as early as 1670 in Tokyo Bay. In autumn of each year, farmers would throw bamboo branches into shallow, muddy water, where the spores of the seaweed would collect. A few weeks later these branches would be moved to a river estuary. The nutrients from the river would help the seaweed to grow.

In the 1940s, the Japanese improved this method by placing nets of synthetic material tied to bamboo poles. This effectively doubled the production. A cheaper variant of this method is called the *hibi* method — simple ropes stretched between bamboo poles.

In the early 1970s there was a recognized demand for seaweed and seaweed products, outstripping supply, and cultivation was viewed as the best means to increase productions.

## Culture Methods

The earliest seaweed farming guides in the Philippines recommended cultivation of *Laminaria* seaweed and reef flats at approximately 1m depth at low tide . They also recommended cutting off sea grasses and removing sea urchins prior to farm construction. Seedlings are then tied to monofilament lines and strung between mangrove stakes pounded into the substrate. This off-bottom method is still one of the major methods used today.

There are new long line cultivation methods that can be used in deeper water approximately 7 m in depth. They use floating cultivation lines anchored to the bottom and are the primary methods used in the villages of North Sulawesi, Indonesia.

Cultivation of seaweed in Asia is a relatively low-technology business with a high labour requirement. There have been many attempts in various countries to introduce high technology to cultivate detached plants growth in tanks on land in order to reduce labour, but they have yet to attain commercial viability.

## Environmental and Ecological Impacts

Several environmental problems can result from seaweed farming. Sometimes seaweed farmers cut down mangroves to use as stakes for their ropes. This, however, negatively affects the farming since it reduces the water quality and mangrove biodiversity due to depletion. Farmers may also sometimes remove eelgrass from their farming areas. This, however, is also discouraged, as it adversely affects water quality.

Seaweed farming helps to preserve coral reefs, by increasing diversity where the algae and seaweed have been introduced and it also provides added niche for local species of fish and invertebrates. Farming may be beneficial by increasing the production of herbivorous fishes and shellfish in the area. Pollnac & et al 1997b reported an increase in Siginid population after the start of extensive farming of eucheuma seaweed in villages in North Sulawesi, Indonesia.

Seaweed culture can also be used to capture, absorb, and eventually incorporate excessive nutrients into living tissue. "Nutrient bioextraction" is the preferred term for bioremediation involving cultured plants and animals. Nutrient bioextraction (also called bioharvesting) is the practice of farming and harvesting shellfish and seaweed for the purpose of removing nitrogen and other nu-trients from natural water bodies.

Seaweed farming can be an actor in biological carbon sequestration.

## Societal Impact

Harvesting seaweed in North Cape (Prince Edward Island)

The practice of seaweed farming has long since spread beyond Japan. In 1997 it was estimated that 40,000 people in the Philippines made their living through seaweed farming. Cultivation is also common in all of southeast Asia, Canada, Great Britain, Spain, and the United States.

## Socioeconomic Aspects

In Japan alone annual production value of nori amounts to US$2 billion and is one of the world's most valuable crops produced by aquaculture. The high demand in seaweed production provides plentiful opportunities and work for the local community. In a study conducted by the Philippines it showed that plots of approximately one hectare can have a net income from eucheuma farming that was 5 to 6 times that of the minimum average wage of an agriculture worker. In the same study they also saw an increase in seaweed exports from 675 metric tons (MT) in 1967 to 13,191 MT in 1980, which doubled to 28,000 MT by 1988.

# Fish Farming

Fish farming or pisciculture involves raising fish commercially in tanks or enclosures, usually for food. It is the principal form of aquaculture, while other methods may fall under mariculture. A facility that releases juvenile fish into the wild for recreational fishing or to supplement a species' natural numbers is generally referred to as a fish hatchery. Worldwide, the most important fish species used in fish farming are carp, tilapia, salmon and catfish.

Koi farming indoors in Israel

Demand is increasing for fish and fish protein, which has resulted in widespread overfishing in wild fisheries. China provides 62 percent of the world's farmed fish. As of 2016, more than 50% of seafood was produced by aquaculture.

Farming carnivorous fish, such as salmon, does not always reduce pressure on wild fisheries, since carnivorous farmed fish are usually fed fishmeal and fish oil extracted from wild forage fish. The 2008 global returns for fish farming recorded by the FAO totaled 33.8 million tonnes worth about $US 60 billion.

Salmon farming in the sea (mariculture) at Loch Ainort, Isle of Skye

## Major Species

| Top 15 cultured fish species by weight in millions of tonnes, according to FAO statistics for 2013 | | | |
|---|---|---|---|
| **Species** | **Environment** | **Tonnage (millions)** | **Value (USD, billion)** |
| Grass carp | freshwater | 5.23 | 6.69 |

| Silver carp | freshwater | 4.59 | 6.13 |
|---|---|---|---|
| Common carp | freshwater | 3.76 | 5.19 |
| Nile tilapia | freshwater | 3.26 | 5.39 |
| Bighead carp | freshwater | 2.90 | 3.72 |
| Catla (Indian carp) | freshwater | 2.76 | 5.49 |
| Crucian carp | freshwater | 2.45 | 2.67 |
| Atlantic salmon | marine | 2.07 | 10.10 |
| Roho labeo | freshwater | 1.57 | 2.54 |
| Milkfish | freshwater | 0.94 | 1.71 |
| Rainbow trout | freshwater, brackish | 0.88 | 3.80 |
| Wuchang bream | freshwater | 0.71 | 1.16 |
| Black carp | freshwater | 0.50 | 1.15 |
| Northern snakehead | freshwater | 0.48 | 0.59 |
| Amur catfish | freshwater | 0.41 | 0.55 |

## Categories

Aquaculture makes use of local photosynthetical production (extensive) or fish that are fed with external food supply (intensive).

## Extensive Aquaculture

Aqua-Boy, a Norwegian live fish carrier used to service the Marine Harvest fish farms on the West coast of Scotland

Growth is limited by available food, commonly zooplankton feeding on pelagic algae or benthic animals, such as crustaceans and mollusks. Tilapia filter feed directly on phytoplankton, which makes higher production possible. Photosynthetic production can be increased by fertilizing pond water with artificial fertilizer mixtures, such as potash, phosphorus, nitrogen and micro-elements.

Another issue is the risk of algal blooms. When temperatures, nutrient supply and available sunlight are optimal for algal growth, algae multiply at an exponential rate, eventually exhausting nutrients and causing a subsequent die-off. The decaying algal biomass will deplete the oxygen in the pond water because it blocks out the sun and pollutes it with organic and inorganic solutes (such as ammonium ions), which can (and frequently do) lead to massive loss of fish.

An alternate option is to use a wetland system such as that of Veta La Palma.

In order to tap all available food sources in the pond, the aquaculturist will choose fish species which occupy different places in the pond ecosystem, e.g., a filter algae feeder such as tilapia, a benthic feeder such as carp or catfish and a zooplankton feeder (various carps) or submerged weeds feeder such as grass carp.

Despite these limitations significant fish farming industries use these methods. In the Czech Republic thousands of natural and semi-natural ponds are harvested each year for trout and carp. The large ponds around Trebon were built from around 1650 and are still in use.

## Intensive Aquaculture

| Optimal water parameters for cold- and warm-water fish in intensive aquaculture | |
|---|---|
| Acidity | pH 6-9 |
| Arsenic | <440 µg/L |
| Alkalinity | >20 mg/L (as $CaCO_3$) |
| Aluminum | <0.075 mg/L |
| Ammonia (non-ionized) | <0.02mg/L |
| Cadmium | <0.0005 mg/L in soft water; < 0.005 mg/L in hard water |
| Calcium | >5 mg/L |
| Carbon dioxide | <5–10 mg/L |
| Chloride | >4.0 mg/L |
| Chlorine | <0.003 mg/L |
| Copper | <0.0006 mg/L in soft water; < 0.03 mg/L in hard water |
| Gas supersaturation | <100% total gas pressure (103% for salmonid eggs/fry) (102% for lake trout) |
| Hydrogen sulfide | <0.003 mg/L |
| Iron | <0.1 mg/L |
| Lead | <0.02 mg/L |
| Mercury | <0.0002 mg/L |
| Nitrate | <1.0 mg/L |
| Nitrite | <0.1 mg/L |
| Oxygen | 6 mg/L for coldwater fish 4 mg/L for warmwater fish |
| Selenium | <0.01 mg/L |
| Total dissolved solids | <200 mg/L |
| Total suspended solids | <80 NTU over ambient levels |
| Zinc | <0.005 mg/L |

In these kinds of systems fish production per unit of surface can be increased at will, as long as sufficient oxygen, fresh water and food are provided. Because of the requirement of sufficient fresh water, a massive water purification system must be integrated in the fish farm. One way to achieve this is to combine hydroponic horticulture and water treatment. The exception to this rule are cages which are placed in a river or sea, which supplements the fish crop with sufficient oxygenated water. Some environmentalists object to this practice.

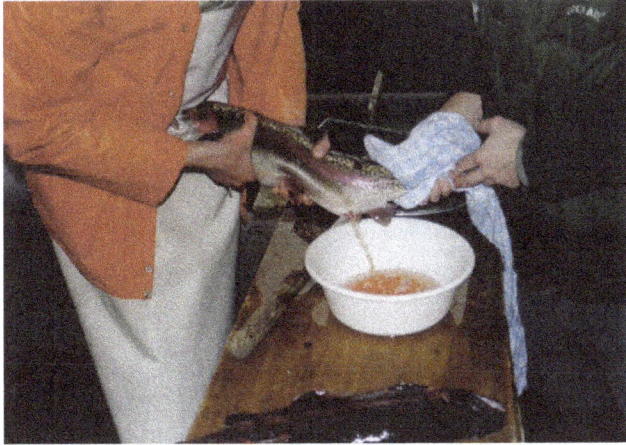

Expressing eggs from a female rainbow trout

The cost of inputs per unit of fish weight is higher than in extensive farming, especially because of the high cost of fish feed, which must contain a much higher level of protein (up to 60%) than cattle food and a balanced amino acid composition as well. However, these higher protein level requirements are a consequence of the higher food conversion efficiency (FCR—kg of feed per kg of animal produced) of aquatic animals. Fish like salmon have an FCR around 1.1 kg of feed per kg of salmon whereas chickens are in the 2.5 kg of feed per kg of chicken range. Fish do not have use energy to keep warm, eliminating a lot of carbohydrates and fats in the diet, required to provide this energy. This however may be offset by the lower land costs and the higher productions which can be obtained due to the high level of input control.

Aeration of the water is essential, as fish need a sufficient oxygen level for growth. This is achieved by bubbling, cascade flow or aqueous oxygen. Catfish, Clarias spp. can breathe atmospheric air and can tolerate much higher levels of pollutants than trout or salmon, which makes aeration and water purification less necessary and makes *Clarias* species especially suited for intensive fish production. In some *Clarias* farms about 10% of the water volume can consist of fish biomass.

The risk of infections by parasites like fish lice, fungi (Saprolegnia spp.), intestinal worms (such as nematodes or trematodes), bacteria (e.g., Yersinia spp., Pseudomonas spp.), and protozoa (such as Dinoflagellates) is similar to animal husbandry, especially at high population densities. However, animal husbandry is a larger and more technologically mature area of human agriculture and better solutions to pathogen problem exist. Intensive aquaculture does have to provide adequate water quality (oxygen, ammonia, nitrite, etc.) levels to minimize stress, which makes the pathogen problem more difficult. This means, intensive aquaculture requires tight monitoring and a high level of expertise of the fish farmer.

Very high intensity recycle aquaculture systems (RAS), where there is control over all the production parameters, are being used for high value species. By recycling the water, very little water is used per unit of production. However, the process does have high capital and operating costs. The higher cost structures mean that RAS is only economical for high value products like broodstock for egg production, fingerlings for net pen aquaculture operations, sturgeon production, research animals and some special niche markets like live fish.

Controlling roes manually

Raising ornamental cold water fish (goldfish or koi), although theoretically much more profitable due to the higher income per weight of fish produced, has never been successfully carried out until very recently. The increased incidences of dangerous viral diseases of koi Carp, together with the high value of the fish has led to initiatives in closed system koi breeding and growing in a number of countries. Today there are a few commercially successful intensive koi growing facilities in the UK, Germany and Israel.

Some producers have adapted their intensive systems in an effort to provide consumers with fish that do not carry dormant forms of viruses and diseases.

In 2016 juvenile Nile tilapia were given a food containing dried *Schizochytrium* in place of fish oil. When compared to a control group raised on regular food, they exhibited higher weight gain and better food-to-growth conversion, plus their flesh was higher in healthy omega-3 fatty acids.

## Fish Farms

Within intensive and extensive aquaculture methods, there are numerous specific types of fish farms; each has benefits and applications unique to its design.

## Cage System

Fish cages are placed in lakes, bayous, ponds, rivers or oceans to contain and protect fish until they can be harvested. The method is also called "off-shore cultivation " when the cages are placed in the sea. They can be constructed of a wide variety of components. Fish are stocked in cages, artificially fed, and harvested when they reach market size. A few advantages of fish farming with cages are that many types of waters can be used (rivers, lakes, filled quarries, etc.), many types of fish can be raised, and fish farming can co-exist with sport fishing and other water uses. Cage farming of fishes in open seas is also gaining popularity. Concerns of disease, poaching, poor water quality, etc., lead some to believe that in general, pond systems are easier to manage and simpler to start. Also, past occurrences of cage-failures leading to escapes, have raised concern regarding the culture of non-native fish species in dam or open-water cages. Even though the cage-industry has made numerous technological advances in cage construction in recent years, storms will always make the concern for escapes valid.

Giant gourami is often raised in cages in central Thailand

Recently, copper alloys have become important netting materials in aquaculture. Copper alloys are antimicrobial, that is, they destroy bacteria, viruses, fungi, algae, and other microbes. In the marine environment, the antimicrobial/algaecidal properties of copper alloys prevent biofouling, which can briefly be described as the undesirable accumulation, adhesion, and growth of microorganisms, plants, algae, tube worms, barnacles, mollusks, and other organisms.

The resistance of organism growth on copper alloy nets also provides a cleaner and healthier environment for farmed fish to grow and thrive. Traditional netting involves regular and labor-intensive cleaning. In addition to its antifouling benefits, copper netting has strong structural and corrosion-resistant properties in marine environments.

Copper-zinc brass alloys are currently (2011) being deployed in commercial-scale aquaculture operations in Asia, South America and the USA (Hawaii). Extensive research, including demonstrations and trials, are currently being implemented on two other copper alloys: copper-nickel and copper-silicon. Each of these alloy types has an inherent ability to reduce biofouling, cage waste, disease, and the need for antibiotics while simultaneously maintaining water circulation and oxygen requirements. Other types of copper alloys are also being considered for research and development in aquaculture operations.

## Irrigation Ditch or Pond Systems

These use irrigation ditches or farm ponds to raise fish. The basic requirement is to have a ditch or pond that retains water, possibly with an above-ground irrigation system (many irrigation systems use buried pipes with headers.)

Using this method, one can store one's water allotment in ponds or ditches, usually lined with bentonite clay. In small systems the fish are often fed commercial fish food, and their waste products can help fertilize the fields. In larger ponds, the pond grows water plants and algae as fish food. Some of the most successful ponds grow introduced strains of plants, as well as introduced strains of fish.

These fish-farming ponds were created as a cooperative project in a rural village.

Control of water quality is crucial. Fertilizing, clarifying and pH control of the water can increase yields substantially, as long as eutrophication is prevented and oxygen levels stay high.Yields can be low if the fish grow ill from electrolyte stress.

## Composite Fish Culture

The Composite fish culture system is a technology developed in India by the Indian Council of Agricultural Research in the 1970s. In this system both local and imported fish species, a combination of five or six fish species is used in a single fish pond. These species are selected so that they do not compete for food among them having different types of food habitats. As a result, the food available in all the parts of the pond is used. Fish used in this system include catla and silver carp which are surface feeders, rohu a column feeder and mrigal and common carp which are bottom feeders. Other fish will also feed on the excreta of the common carp and this helps contribute to the efficiency of the system which in optimal conditions will produce 3000–6000 kg of fish per hectare per year.

One problem with such composite fish culture is that many of these fish breed only during monsoon. Even if fish seed is collected from the wild, it can be mixed with that of other species as well. So, a major problem in fish farming is the lack of availability of good-quality seed. To overcome this problem, ways have now been worked out to breed these fish in ponds using hormonal stimulation. This has ensured the supply of pure fish seed in desired quantities.

## Integrated Recycling Systems

One of the largest problems with freshwater pisciculture is that it can use a million gallons of water per acre (about 1 m$^3$ of water per m$^2$) each year. Extended water purification systems allow for the reuse (recycling) of local water.

The largest-scale pure fish farms use a system derived (admittedly much refined) from the New Alchemy Institute in the 1970s. Basically, large plastic fish tanks are placed in a greenhouse. A hydroponic bed is placed near, above or between them. When tilapia are raised in the tanks, they are able to eat algae, which naturally grow in the tanks when the tanks are properly fertilized.

The tank water is slowly circulated to the hydroponic beds where the tilapia waste feeds commercial plant crops. Carefully cultured microorganisms in the hydroponic bed convert ammonia to nitrates, and the plants are fertilized by the nitrates and phosphates. Other wastes are strained out by the hydroponic media, which doubles as an aerated pebble-bed filter.

This system, properly tuned, produces more edible protein per unit area than any other. A wide variety of plants can grow well in the hydroponic beds. Most growers concentrate on herbs (e.g. parsley and basil), which command premium prices in small quantities all year long. The most common customers are restaurant wholesalers.

Since the system lives in a greenhouse, it adapts to almost all temperate climates, and may also adapt to tropical climates. The main environmental impact is discharge of water that must be salted to maintain the fishes' electrolyte balance. Current growers use a variety of proprietary tricks to keep fish healthy, reducing their expenses for salt and waste water discharge permits. Some veterinary authorities speculate that ultraviolet ozone disinfectant systems (widely used for ornamental fish) may play a prominent part in keeping the Tilapia healthy with recirculated water.

A number of large, well-capitalized ventures in this area have failed. Managing both the biology and markets is complicated. One future development is the combination of Integrated Recycling systems with Urban Farming as tried in Sweden by the Greenfish initiative.

## Classic Fry Farming

This is also called a "Flow through system" Trout and other sport fish are often raised from eggs to fry or fingerlings and then trucked to streams and released. Normally, the fry are raised in long, shallow concrete tanks, fed with fresh stream water. The fry receive commercial fish food in pellets. While not as efficient as the New Alchemists' method, it is also far simpler, and has been used for many years to stock streams with sport fish. European eel (Anguilla anguilla) aquaculturalists procure a limited supply of glass eels, juvenile stages of the European eel which swim north from the Sargasso Sea breeding grounds, for their farms. The European eel is threatened with extinction because of the excessive catch of glass eels by Spanish fishermen and overfishing of adult eels in, e.g., the Dutch IJsselmeer, Netherlands. As per 2005, no one has managed to breed the European eel in captivity.

## Issues

The issue of feeds in fish farming has been a controversial one. Many cultured fishes (tilapia, carp, catfish, many others) require no meat or fish products in their diets. Top-level carnivores (most salmon species) depend on fish feed of which a portion is usually derived from wild caught (anchovies, menhaden, etc.). Vegetable-derived proteins have successfully replaced fish meal in feeds for carnivorous fishes, but vegetable-derived oils have not successfully been incorporated into the diets of carnivores.

Secondly, farmed fish are kept in concentrations never seen in the wild (e.g. 50,000 fish in a 2-acre (8,100 m²) area.). However, fish tend also to be animals that aggregate into large schools at high density. Most successful aquaculture species are schooling species, which do not have social problems at high density. Aquaculturists tend to feel that operating a rearing system above its design capacity or above the social density limit of the fish will result in decreased growth rate and increased FCR (food conversion ratio - kg dry feed/kg of fish produced), which will result in increased cost and risk of health problems along with a decrease in profits. Stressing the animals is not desirable, but the concept of and measurement of stress must be viewed from the perspective of the animal using the scientific method.

Sea lice, particularly *Lepeophtheirus salmonis* and various *Caligus* species, including *Caligus clemensi* and *Caligus rogercresseyi*, can cause deadly infestations of both farm-grown and wild salmon. Sea lice are ectoparasites which feed on mucus, blood, and skin, and migrate and latch onto the skin of wild salmon during free-swimming, planktonic *nauplii* and *copepodid* larval stages, which can persist for several days. Large numbers of highly populated, open-net salmon farms can create exceptionally large concentrations of sea lice; when exposed in river estuaries containing large numbers of open-net farms, many young wild salmon are infected, and do not survive as a result. Adult salmon may survive otherwise critical numbers of sea lice, but small, thin-skinned juvenile salmon migrating to sea are highly vulnerable. On the Pacific coast of Canada, the louse-induced mortality of pink salmon in some regions is commonly over 80%.

A 2008 meta-analysis of available data shows that salmon farming reduces the survival of associated wild salmon populations. This relationship has been shown to hold for Atlantic, steelhead, pink, chum, and coho salmon. The decrease in survival or abundance often exceeds 50 percent.

Diseases and parasites are the most commonly cited reasons for such decreases. Some species of sea lice have been noted to target farmed coho and Atlantic salmon. Such parasites have been shown to have an effect on nearby wild fish. One place that has garnered international media attention is British Columbia's Broughton Archipelago. There, juvenile wild salmon must "run a gauntlet" of large fish farms located off-shore near river outlets before making their way to sea. It is alleged that the farms cause such severe sea lice infestations that one study predicted in 2007 a 99% collapse in the wild salmon population by 2011. This claim, however, has been criticized by numerous scientists who question the correlation between increased fish farming and increases in sea lice infestation among wild salmon.

Because of parasite problems, some aquaculture operators frequently use strong antibiotic drugs to keep the fish alive (but many fish still die prematurely at rates of up to 30 percent). In some cases, these drugs have entered the environment. Additionally, the residual presence of these drugs in human food products has become controversial. Use of antibiotics in food production is thought to increase the prevalence of antibiotic resistance in human diseases. At some facilities, the use of antibiotic drugs in aquaculture has decreased considerably due to vaccinations and other techniques. However, most fish farming operations still use antibiotics, many of which escape into the surrounding environment.

The lice and pathogen problems of the 1990s facilitated the development of current treatment methods for sea lice and pathogens. These developments reduced the stress from parasite/pathogen problems. However, being in an ocean environment, the transfer of disease organisms from the wild fish to the aquaculture fish is an ever-present risk.

The very large number of fish kept long-term in a single location contributes to habitat destruction of the nearby areas. The high concentrations of fish produce a significant amount of condensed faeces, often contaminated with drugs, which again affect local waterways. However, if the farm is correctly placed in an area with a strong current, the 'pollutants' are flushed out of the area fairly quickly. Not only does this help with the pollution problem. but water with a stronger current also aids in overall fish growth. Concern remains that resultant bacterial growth strips the water of oxygen, reducing or killing off the local marine life. Once an area has been so contaminated, the fish farms are moved to new, uncontaminated areas. This practice has angered nearby fishermen.

Other potential problems faced by aquaculturists are the obtaining of various permits and water-use rights, profitability, concerns about invasive species and genetic engineering depending on what species are involved, and interaction with the United Nations Convention on the Law of the Sea.

In regards to genetically modified farmed salmon, concern has been raised over their proven reproductive advantage and how it could potentially decimate local fish populations, if released into the wild. Biologist Rick Howard did a controlled laboratory study where wild fish and GMO fish were allowed to breed. The GMO fish crowded out the wild fish in spawning beds, but the offspring were less likely to survive. The colorant used to make pen-raised salmon appear rosy like their wild cousins has been linked with retinal problems in humans.

## Labeling

In 2005, Alaska passed legislation requiring that any genetically altered fish sold in the state be labeled. In 2006, a *Consumer Reports* investigation revealed that farm-raised salmon is frequently sold as wild.

In 2008, the US National Organic Standards Board allowed farmed fish to be labeled as organic provided less than 25% of their feed came from wild fish. This decision was criticized by the advocacy group Food & Water Watch as "bending the rules" about organic labeling. In the European Union, fish labeling as to species, method of production and origin, has been required since 2002.

Concerns continue over the labeling of salmon as farmed or wild caught, as well as about the humane treatment of farmed fish. The Marine Stewardship Council has established an Eco label to distinguish between farmed and wild caught salmon, while the RSPCA has established the Freedom Food label to indicate humane treatment of farmed salmon as well as other food products.

## Indoor Fish Farming

An alternative to outdoor open ocean cage aquaculture, is through the use of a recirculation aquaculture system (RAS). A RAS is a series of culture tanks and filters where water is continuously recycled and monitored to keep optimal conditions year round. To prevent the deterioration of water quality, the water is treated mechanically through the removal of particulate matter and biologically through the conversion of harmful accumulated chemicals into non-toxic ones.

Other treatments such as UV sterilization, ozonation, and oxygen injection are also used to maintain optimal water quality. Through this system, many of the environmental drawbacks of aquaculture are minimized including escaped fish, water usage, and the introduction of pollutants. The practices also increased feed-use efficiency growth by providing optimum water quality (Timmons et al., 2002; Piedrahita, 2003).

One of the drawbacks to recirculation aquaculture systems is water exchange. However, the rate of water exchange can be reduced through aquaponics, such as the incorporation of hydroponically grown plants (Corpron and Armstrong, 1983) and denitrification (Klas et al., 2006). Both methods reduce the amount of nitrate in the water, and can potentially eliminate the need for water exchanges, closing the aquaculture system from the environment. The amount of interaction between the aquaculture system and the environment can be measured through the cumulative feed burden (CFB kg/M3), which measures the amount of feed that goes into the RAS relative to the amount of water and waste discharged.

From 2011, a team from the University of Waterloo led by Tahbit Chowdhury and Gordon Graff examined vertical RAS aquaculture designs aimed at producing protein-rich fish species. However, because of its high capital and operating costs, RAS has generally been restricted to practices such as broodstock maturation, larval rearing, fingerling production, research animal production, SPF (specific pathogen free) animal production, and caviar and ornamental fish production. As such, research and design work by Chowdhury and Graff remains difficult to implement. Although the use of RAS for other species is considered by many aquaculturalists to be currently impractical, there has been some limited successful implementation of this with high value product such as barramundi, sturgeon and live tilapia in the US eels and catfish in the Netherlands, trout in Denmark and salmon is planned in Scotland and Canada.

## Slaughter Methods

Tanks saturated with carbon dioxide have been used to make fish unconscious. Their gills are then cut with a knife so that the fish bleed out before they are further processed. This is no longer considered a humane method of slaughter. Methods that induce much less physiological stress are electrical or percussive stunning and this has led to the phasing out of the carbon dioxide slaughter method in Europe.

## Inhumane Methods

According to T. Håstein of the National Veterinary Institute, "Different methods for slaughter of fish are in place and it is no doubt that many of them may be considered as appalling from an animal welfare point of view." A 2004 report by the EFSA Scientific Panel on Animal Health and Welfare explained: "Many existing commercial killing methods expose fish to substantial suffering over a prolonged period of time. For some species, existing methods, whilst capable of killing fish humanely, are not doing so because operators don't have the knowledge to evaluate them." Following are some of the less humane ways of killing fish.

- *Air Asphyxiation.* This amounts to suffocation in the open air. The process can take upwards of 15 minutes to induce death, although unconsciousness typically sets in sooner.

- *Ice baths / chilling.* Farmed fish are sometimes chilled on ice or submerged in near-freezing water. The purpose is to dampen muscle movements by the fish and to delay the onset of post-death decay. However, it does not necessarily reduce sensibility to pain; indeed, the chilling process has been shown to elevate cortisol. In addition, reduced body temperature extends the time before fish lose consciousness.

- *CO2 narcosis.*

- *Exsanguination without stunning.* This is a process in which fish are taken up from water, held still, and cut so as to cause bleeding. According to references in Yue, this can leave fish writhing for an average of four minutes, and some catfish still responded to noxious stimuli after more than 15 minutes.

- *Immersion in salt followed by gutting or other processing such as smoking.* This process is applied to eel.

## More Humane Methods

Proper stunning renders the fish unconscious immediately and for a sufficient period of time such that the fish is killed in the slaughter process (e.g. through exsanguination) without regaining consciousness.

- *Percussive stunning.* This involves rendering the fish unconscious with a blow on the head.

- *Electric stunning.* This can be humane when a proper current is made to flow through the fish brain for a sufficient period of time. Electric stunning can be applied after the fish has been taken out of the water (dry stunning) or while the fish is still in the water. The latter generally requires a much higher current and may lead to operator safety issues. An advantage could be that in-water stunning allows fish to be rendered unconscious without stressful handling or displacement. However, improper stunning may not induce insensibility long enough to prevent the fish from enduring exsanguination while conscious. It's unknown whether the optimal stunning parameters that researchers have determined in studies are used by the industry in practice.

## Photo Gallery

Houseboat rafts with cages under for rearing fish. Near My Tho, Vietnam

Transport boats moored at fish processing plant. My Tho, Vietnam

Communal Zapotec fish farm in Ixtlán de Juárez, Mexico

# Geoduck Aquaculture

Geoduck aquaculture or geoduck farming is the practice of cultivating geoducks (specifically the Pacific geoduck, Panopea generosa) for human consumption. The geoduck is a large edible saltwater clam, a marine bivalve mollusk, that is native to the Pacific Northwest.

Juvenile geoducks are planted or seeded on the ocean floor or substrate within the soft intertidal and subtidal zones, then harvested five to seven years later when they have reached marketable size (about 1 kg or 2.2 lbs). They are native to the Pacific region and are found from Baja California, through the Pacific Northwest and Southern Alaska.

Most geoducks are harvested from the wild, but because of state government-instituted limits on the amount that can be harvested, the need to grow geoducks in farms to meet an increasing demand has led to the growth of the geoduck aquaculture industry, particularly in Puget Sound, Washington. Geoduck meat is a prized delicacy in Asian cuisine; the majority of exports are sent to China (Shanghai, Shenzhen, Guangzhou, Beijing, are the main Chinese markets), Hong Kong and Japan.

Geoducks on display as seafood in a Chinese restaurant in Hong Kong

# History

## 1. Washington State

Wild geoducks had been harvested in Puget Sound, Washington by residents and visitors for hundreds of years, but it was not until 1970 that the Washington Department of Natural Resources (WDNR) auctioned off the first right to commercially harvest wild geoducks. Research into the viability of farming geoducks began in the 1970s. In 1991, the development of hatchery and grow-out methods from brood stock were initiated. By 1996, commercial aquaculture had begun. As of 2011, there were 237 commercial sites operating on 145 ha of privately owned properties (including those leased from other private owners). Commercial geoduck aquaculture has been primarily undertaken within the intertidal zone.

## 2. British Columbia

Commercial harvesting of wild geoducks began in 1976. In the early 1990s, the cultivation method developed in Washington was adopted in British Columbia by Fan Seafoods Ltd and the Underwater Harvesters Association (UHA), a group of 55 licence holders for geoduck and horse clam fishery. The UHA used this method to initiate a wild geoduck enhancement program by seeding depleted subtidal areas with cultivated juvenile geoducks thereby ensuring continued supply in the wild. It even invented a mechanical seeder that plants cultured juvenile geoducks on subtidal beds. Through a collaboration agreement between the provincial government's Department of Fisheries and Oceans (DFO), Fan Seafoods Ltd. and UHA, five pilot sites were selected in 1996 to study the feasibility of a geoduck aquaculture venture. In 2007, the provincial government of B.C. licensed UHA to operate the first commercial geoduck farm on 25.3 ha off Hernando Island.

## 3. Other Areas

No geoduck aquaculture industry exists in Southern Alaska and Mexico. In New Zealand, Cawthron recently reported successful attempts at rearing juvenile geoducks. The plan is to plant them in subtidal areas in order to supplement wild geoduck harvest.

## Geoduck Species and their Distribution

*Panopea generosa* is the geoduck species that is found in the Pacific Northwest and Alaska. *Panopea globosa*, which is another species in the same genus, *Panopea*, is harvested in Mexico's Gulf of California.

A small wild geoduck fishery exists in New Zealand for *Panopea zelandica*, the "deepwater clam", and in Argentina for *Panopea abbreviata*, the "southern geoduck". A fifth species, *Panopea japonica*, the Japanese geoduck, is found in Korea and Japan, but there is no viable commercial industry in those countries for this species.

Biomass densities in Southeast Alaska are estimated by divers then inflated by twenty percent to account for geoducks not visible at the time of survey. This estimate is used to predict the two percent allowed for commercial harvesting.

## Predators and Diseases

Juvenile geoducks are susceptible to attack from predators in their first year when they have not yet burrowed deeply into the substrate. Crabs, sea stars, predatory gastropods, and flatfishes have been observed to feed on them. Adult geoducks, which are already buried deep in the substrate, are out of reach of most predators except for sea otters and humans.

In 2012, no infectious diseases had been observed attacking cultured juvenile geoducks planted in the wild up to that point. Surface abnormalities were observed in wild adult geoducks, but the pathogen or pathogens could not be identified. However, a protozoan parasite (*Isonema* sp) was believed to be the causal agent of cultured geoduck larvae mortalities at a Washington State experimental hatchery.

## Production Methods

The Washington Department of Fisheries (WDF) Point Whitney Laboratory pioneered research into the aquaculture of geoducks in 1970. The initial purpose of developing the techniques was to enhance the wild population that was being depleted by commercial fishing. Their first challenge was inducing spawning from wild adult geoducks brought into the hatchery; the second challenge was the survival of the resulting larvae. As per 2012, research into improving culture techniques is continuing, however the basic environmental conditions for growth of geoducks have already been established.

## Table 1. Summary of Optimal Biophysical Parameters for Geoduck Culture (Nursery and Grow-out)

| Parameter | Optimal Value |
|-----------|---------------|
| Substrate | mud/sand/pea gravel (penetration to 1m) |
| Depth | 3–20 m |

| Temperature | 8-18 C |
|---|---|
| Salinity | 26-31 ppt |
| Transparency (Secchi) | 2->10m |
| Current velocity | <1.5 kn (<0.75 cm/s) |
| Productivity | 15-200 mgC/m2/day |

The techniques for culturing geoducks is similar to that of other bivalves. Modifications have been made by both academic and private laboratories through the years.

## Collection of Broodstock

Geoducks spawn from spring to late summer in the wild, peaking in June and July. Because of this timing, an equal number of male and female clams are collected starting in the early fall when gametogenesis commences. The clams are placed in milk crates and maintained in polyethylene fish totes supplied with flowing seawater (10-12 °C) for several weeks. Microalgae is added as feed and regular cleaning is carried out to remove biodeposits.

## Spawning Method

Spawning is initiated by changing the seawater and increasing the amount of microalgae to increase the temperature of the water. The higher temperature and abundant supply of microalgal feed induces spawning in males first, then in females.

## Rearing of Larvae

Fertilized geoduck eggs remain floating in the water column for 16 to 35 days until they metamorphose and settle on the substrate. As larvae, they are kept at a water temperature of 16 °C and supplied with microalgal feed with frequent water changes.

## Growing-out Method

Larvae that are ready to metamorphose are collected and placed in a primary nursery system where water temperature is kept at 15-17 °C and supplied with microalgal feed. Metamorphosed larvae are characterized by the development of an attachment mechanism known as byssal threads.

Once byssal threads have developed, the clams are moved to a secondary nursery system which contains sand as the substrate. They are kept here until they are large enough to be moved outdoors.

Tertiary nursery systems are made of large outdoor tanks or totes which have the same sand substrate and flowing seawater. The clams are kept in these systems until they reach a valve length of 5 mm, at which point they are ready to be planted.

## Planting and Seeding

Four to five juvenile geoducks are planted inside PVC tubes that are "wiggled" into the sandy substrate along the intertidal zone during low tide. The PVC tubes are between 5 and 15 cm in diameter, with lengths from 20 to 30 cm, about 7 cm of which remain above the substrate. The plastic tubes are covered with a mesh net to protect the clams from predators. The tubes also serve to retain seawater at low tide, which prevents dehydration of the clams. After one to two growing seasons when the juvenile geoducks have burrowed themselves deep enough into the substrate to be out of reach of predators, the PVC tubes are removed. Not all tidelands are suitable for geoduck aquaculture. The sand must be deep and clean, and the water must have the right salinity and degree of cleanliness.

It should be noted that in Washington State, the aquaculture of geoducks occurs on intertidal lands, whereas in British Columbia, geoducks are cultured in subtidal areas, which necessitates the growing of juvenile geoducks to at least 12 mm instead of 5 mm before planting. Once planted in the subtidal bed, the area is covered with netting to protect the clams from predators (PVC tubes are not stable in subtidal beds due to strong currents).

## Harvesting

Mature geoducks are left to grow out until they are large enough to be marketable (1.0 kg). This can take from five to seven years. Wild and cultured geoducks are harvested by first loosening the substrate around them using a powerful nozzle that ejects high-pressured water. Once loosened, the clams are collected by hand and placed in crates for transport to a processing facility.

## Grading and Packing

Although there is no standard grading system for quality, the color of the siphon (the whiter the better) and the size (up to 1 kg) are the main determiners of price. Live geoducks are packed in coolers and shipped on the same day they are harvested.

## Economic Importance

The geoduck industry produces an estimated 6000 metric tons of clams annually, of which only about 10-13% come from aquaculture. Washington is the largest producer of wild and cultured geoducks.

## Table 2. Average Annual Production 2007-2010

| | Location | Production (in metric tons) |
|---|---|---|
| Wild | Washington State | 2143 |
| | British Columbia | 1572 |
| | Mexico | 1094 |
| Culture | Washington State | 591 |
| | British Columbia | 75 |
| Total | | 5823 |

# Table 3. Difference between Wild and Cultured Geoduck

| Wild | Culture |
|---|---|
| crunchier | generally whiter |
| hardier, travels better | more uniform size |
| less water shrinkage | thinner shell |
|  | may weigh 25% less than average wild geoduck |

## Major Markets

China (Mainland and Hong Kong) is where 95% of geoducks exports are sent. Although the clams are priced at about $20 per pound at the point of origin, they can sell for $100 to $150 per pound at their destination. While exports to Japan have decreased in recent years because of increasing prices, the market in China is expected to soar.

## Management of the Industry

### British Columbia

Environmental groups and citizens of British Columbia have voiced their concerns about geoduck aquaculture operations in the Province, even though the industry is still in its preliminary stages. Their main issue has been the lack of peer-reviewed studies on the impact geoduck aquaculture practices will have on the environment. Most concerned groups point to the situation in Puget Sound, Washington as an example of the environmental harm posed by geoduck farms. Other concerns being raised include the destruction of the natural aquatic habitat, washed up waste (such as nets), disease outbreaks, competition with wild species, and "purge fishing" or the removable of all wild geoducks in a specific area prior to the planting of cultured geoducks. This procedure is apparently necessary because of the economics of the industry.

The management of Canada's aquaculture sector is headed by the Department of Fisheries and Oceans. The department shares this responsibility with 17 other departments and agencies at the federal and provincial levels. The DFO works with these government offices to "create the policy and regulatory conditions necessary to ensure that the aquaculture industry develops in an environmentally responsible way while remaining economically competitive in national and international markets". In the case of wild geoduck fishery, the agency co-manages the activity with the Underwater Harvesters Association.

Aquaculture in Canada is regulated by three main acts: the Fisheries Act, Navigable Waters Protection Act, and Canadian Environmental Assessment Act. Other acts that control aquaculture practices include the Land Act, Health of Animals Act, Food and Drugs Act, Pest Control Products Act, and Species at Risk Act. All of these acts specify regulations at the local, state and federal levels, resulting in a total of 73 rules and regulations for the aquaculture industry; these rules and regulations have been described as being conflicting and contradictory. The rules and regulations have resulted in the aquaculture industry being described as "one of the most heavily regulated in the world". A recent survey showed that Canadians support the creation of an Aquaculture Act that specifically addresses the needs of the industry. The DFO collects fees from aquaculture licences and leases, and receives government funding for its research programs. The UHA also funds research on geoduck aquaculture.

To address consumer concerns regarding unsafe aquaculture practices, the DFO launched the Aquaculture Sustainability Reporting Initiative in 2011. This report backs the Federal Sustainable Development Strategy implemented in 2010, and aims to provide its citizens with information on the sustainable aquaculture practices that government agencies and the aquaculture industry are undertaking or plan to undertake. There are currently 29 participants in this initiative, coming from different sectors such as academia, the aquaculture industry, government agencies, and environmental organizations.

The Canadian government and the aquaculture industry demonstrate sustainable practices by several means such as federal (Canadian General Standards Board) and third-party certifications (International Organization for Standardization for traceability of produce). The Aquaculture Sustainability Reporting Initiative is patterned after the Global Reporting Initiative, which emphasizes reporting transparency and the accountability of an organization's sustainability performance.

The DFO also recently released Aquaculture in Canada 2012: A Report on Aquaculture Sustainability in which it outlines its performance in terms of sustainability. The aquaculture industry has also taken steps to develop a Codes of Practice for sustainable operations that are in line with or exceed international standards. In the case of geoduck, the UHA has adopted a labeling system ("Market Approved") to ensure that the geoducks that end up in the market are safe to eat, of approved quality, and not illegally harvested.

The DFO plans to undertake geoduck aquaculture in subtidal areas. No geoduck production currently occurs on private tidelands, although conversion of other shellfish aquaculture ventures operating on tidelands to geoduck is also being considered. There is no large-scale commercial production underway yet; ongoing trial farms are currently being studied and assessed. Although tenures to possible geoduck farm sites have been granted, commercial licences have not been issued, except for the one granted to the UHA.

Marketing and promotion When promoting its products, the Canadian aquaculture industry touts the environmentally sound practices it observes in producing high-quality fish and shellfish. The Canadian Aquaculture Industry Alliance recently received backing from the DFO with a $1 M investment to promote awareness of the industry and increase sales. The Aquaculture Innovation and Market Access Program (AIMAP) of the DFO aims to encourage technological innovation in the industry to improve its "global competitiveness and environmental performance". However, in the case of geoduck there is no formal marketing and promotion underway. Since its main market is China, this industry has relied on connections between Vancouver-based export businesses with close ties (especially familial ties) to Hong Kong and mainland China importers. The UHA however has been promoting geoducks in China with support from the federal government.

## Washington State

Concerns have also been raised regarding the impact of geoduck aquaculture on the natural habitat, particularly in Puget Sound. Currently, geoduck aquaculture in Puget Sound occupies 80 ha of private tidelands which are either owned by aquaculture companies or leased from other landowners. (another report put the area at 141 ha) Because geoduck aquaculture occurs on private lands, there is minimal government oversight, and environmental concerns raised by the residents are most often left to the aquaculture companies to address, and in some cases for the courts to arbitrate. The aquaculture companies do create their own environmental codes of conduct and best management practices to address such concerns.

The state government is considering leasing public aquatic lands (state-owned) specifically for geoduck aquaculture. It currently leases 849 ha for aquaculture of other shellfish, such as oysters, other kinds of clams, and mussels. Fees are collected from aquaculture companies, and the resulting revenue is used to manage and protect public aquatic lands throughout the State. Since its statehood in 1889, Washington had been selling tidelands to private individuals, initially as a source of revenue for the state. By 1971, when this practice was stopped, the State had already sold about 60% of public tidelands to private ownership. The state currently owns 1 million ha of aquatic lands.

Several state aquatic land statutes enacted under the Shoreline Management Act of 1971 gave authority to the DNR to "foster the commercial and recreational use of the aquatic environment for production of food, fibre, income, and public enjoyment from state-owned aquatic lands under its jurisdiction and from associated waters, and to this end the department may develop and improve production and harvesting of seaweeds and sealife attached to or growing on aquatic land or contained in aquaculture containers..."

Aquaculture is given priority in Washington: "The legislature finds that many areas of the state of Washington are scientifically and biologically suitable for aquaculture development, and therefore the legislature encourages promotion of aquacultural activities, programs, and development with the same status as other agricultural activities, programs, and development within the state". At the national level in the USA, the National Oceanic and Atmospheric Administration (NOAA) is the lead agency for aquaculture. In February 2011, this agency released a draft of the National Policy for Sustainable Marine Aquaculture in the United States of America that aims to protect but, at the same time, utilize the nation's aquatic resources in a sustainable manner as well as encourage the growth of a sustainable aquaculture industry.

Commercial aquaculture in Washington is regulated by local, state, and federal government entities, each tasked with different responsibilities. Some of the agencies involved are the Environmental Protection Agency, Department of Fisheries and Wildlife, US Army Corps of Engineers, and the Food and Drug Administration. The decisions of these agencies are governed by several federal acts, such as the Clean Water Act, Lacey Act, Federal Water Pollution Control Act, and Animal Health Protection Act.

Because of the concerns raised by residents and environmental groups regarding the ecological impact of geoduck aquaculture on private tidelands, the WDNR has adopted a more cautious approach on leasing state-owned aquatic lands for geoduck aquaculture. In 2003, the State legislature instructed the WDNR to explore the feasibility of a geoduck aquaculture program on state-owned tidelands. In 2007, the state passed House Bill 2220 on Shellfish Aquaculture which, among other things, commissions the Washington Sea Grant (WSG) of the University of Washington to conduct "a series of scientific research studies that examines the possible effects, including the cumulative effects, of the current prevalent geoduck aquaculture techniques and practices on the natural environment in and around Puget Sound, including the Strait of Juan de Fuca". The research is expected to end on December 1, 2013. The bill further stipulates that not more than 15 ha of state-owned aquatic land be leased for commercial geoduck aquaculture every year until 2014.

It also created the Shellfish Aquaculture Regulatory Committee, which is composed of government agencies, aquaculture producers (2), concerned environmental organizations (2), and landowners (2). The role of the committee is to recommend guidelines and policies for shellfish aquaculture operations. In 2010, the WDNR tok a further step further by opening a dialogue with stakeholders and the public. They created an online forum on geoduck aquaculture to elicit concerns from residents, environmental groups and geoduck farm owners.

Marketing and promotion Half of the geoducks produced in Washington is exported to Vancouver, BC. before being re-exported to the final markets in China and Hong Kong. The remaining half are exported through Seattle, WA and Anchorage, AK. These three cities have the best air connections to China and Hong Kong. Even though China is Washington's biggest market for geoduck, there is little promotion from the state's geoduck producers there.

## Washington Sea Grant Studies

In order to address priorities set by the Washington State legislature, the WSG is conducting research on three key areas:

1. Geochemical and Ecological Consequences of Disturbances Associated with Geoduck Aquaculture Operations in Washington.

2. Cultured-Wild Interactions: Disease Prevalence in Wild Geoduck Populations.

3. Resilience of Soft-Sediment Communities after Geoduck Harvest in Samish Bay, Washington State.

The WSG released its most recent progress report in February 2012 on the possible effects of geoduck aquaculture on the environment. The preliminary results of some of the studies appear to show that geoduck aquaculture does not negatively affect the natural habitat. One of the studies have been completed, and results showed that the seemingly disruptive nature of harvesting geoducks has no effect on the infaunal benthic community. The report suggested that because the infauna are already accustomed to natural disturbances such as wave action and extreme weather conditions, harvesting does not affect them any differently. This report, however, has garnered criticisms which point out that the studies are not long-term, so the effect of geoduck aquaculture practices over many years still cannot be ascertained.

## Freshwater Prawn Farming

A freshwater prawn farm is an aquaculture business designed to raise and produce freshwater prawns or shrimp[1] for human consumption. Freshwater prawn farming shares many characteristics with, and many of the same problems as, marine shrimp farming. Unique problems are introduced by the developmental life cycle of the main species (the giant river prawn, *Macrobrachium rosenbergii*).

The global annual production of freshwater prawns (excluding crayfish and crabs) in 2003 was about 280,000 tons, of which China produced some 180,000 tons, followed by India and Thailand with some 35,000 tons each. Additionally, China produced about 370,000 tons of Chinese river crab (*Eriocheir sinensis*).

A farmer constructing a shrimp farm in Pekalongan, Indonesia

## Species

All farmed freshwater prawns today belong to the genus *Macrobrachium*. Until 2000, the only species farmed was the giant river prawn (*Macrobrachium rosenbergii*, also known as the Malaysian prawn). Since then, China has begun farming the Oriental river prawn (*M. nipponense*) in large quantities, and India farms a small amount of monsoon river prawn (*M. malcolmsonii*). In 2003, these three species accounted for all farmed freshwater prawns, about two-thirds *M. rosenbergii* and one-third *M. nipponense*.

About 200 species in the genus *Macrobrachium* live in the tropical and subtropical climates on all continents except Europe and Antarctica.

Biology of *Macrobrachium rosenbergii*

Giant river prawns live in turbid freshwater, but their larval stages require brackish water to survive. Males can reach a body size of 32 cm; females grow to 25 cm. In mating, the male deposits spermatophores on the underside of the female's thorax, between the walking legs. The female then extrudes eggs, which pass through the spermatophores. The female carries the fertilized eggs with her until they hatch; the time may vary, but is generally less than three weeks. A large female may lay up to 100,000 eggs.

From these eggs hatch zoeae, the first larval stage of crustaceans. They go through several larval stages before metamorphosing into postlarvae, at which stage they are about 8 mm long and have all the characteristics of adults. This metamorphosis usually takes place about 32 to 35 days after hatching. These postlarvae then migrate back into freshwater.

There are three different morphotypes of males. The first stage is called "small male" (SM); this smallest stage has short, nearly translucent claws. If conditions allow, small males grow and metamorphose into "orange claw" (OC) males, which have large orange claws on their second chelipeds, which may have a length of 0.8 to 1.4 times their body size. OC males later may transform into the third and final stage, the "blue claw" (BC) males. These have blue claws, and their second chelipeds may become twice as long as their body.

Male *M. rosenbergii* prawns have a strict hierarchy: the territorial BC males dominate the OCs, which in turn dominate the SMs. The presence of BC males inhibits the growth of SMs and delays the metamorphosis of OCs into BCs; an OC will keep growing until it is larger than the largest BC male in its neighbourhood before transforming. All three male stages are sexually active, though, and females which have undergone their premating molt will cooperate with any male to reproduce. BC males protect the females until their shells have hardened; OCs and SMs show no such behavior.

## Technology

Giant river prawns have been farmed using traditional methods in Southeast Asia for a long time. First experiments with artificial breeding cultures of *M. rosenbergii* were done in the early 1960s in Malaysia, where it was discovered that the larvae needed brackish water for survival. Industrial-scale rearing processes were perfected in the early 1970s in Hawaii, and spread first to Taiwan and Thailand, and then to other countries.

The technologies used in freshwater prawn farming are basically the same as in marine shrimp farming. Hatcheries produce postlarvae, which then are grown and acclimated in nurseries before being transferred into growout ponds, where the prawns are then fed and grown until they reach marketable size. Harvesting is done by either draining the pond and collecting the animals ("batch" harvesting) or by fishing the prawns out of the pond using nets (continuous operation).

Due to the aggressive nature of *M. rosenbergii* and the hierarchy between males, stocking densities are much lower than in penaeid shrimp farms. Intensive farming is not possible due to the increased level of cannibalism, so all farms are either stocked semi-intensively (4 to 20 postlarvae per square metre) or, in extensive farms, at even lower densities (1 to 4/m²). The management of the growout ponds must take into account the growth characteristics of *M. rosenbergii*: the presence of blue-claw males inhibits the growth of small males, and delays the metamorphosis of OC males into BCs. Some farms fish off the largest prawns from the pond using seines to ensure a healthy composition of the pond's population, designed to optimize the yield, even if they employ batch harvesting. The heterogeneous individual growth of *M. rosenbergii* makes growth control necessary even if a pond is stocked newly, starting from scratch: some animals will grow faster than others and become dominant BCs, stunting the growth of other individuals.

Freshwater prawn farm in Bangladesh.

The FAO considers the ecological impact of freshwater prawn farming to be less severe than in shrimp farming. The prawns are cultured at much lower densities, meaning less concentrated waste products and a lesser danger of the ponds becoming breeding places for diseases. The grow-out ponds do not salinate agricultural land, as do those of inland marine shrimp farms. However, the lower yield per area means that the income per Ha is also lower and a given area can support fewer humans. This limits the culture area to low value lands where intensification is not required. Freshwater prawn farms do not endanger mangroves, and are better amenable to small-scale businesses run by a family. However, like marine farmed shrimp, *M. rosenbergii* is also susceptible to a variety of viral or bacterial diseases, including white tail disease, also called "white muscle disease".

## Economics

The global annual production of freshwater prawns in 2003 was about 280,000 tonnes, of which China produced some 180,000 tonnes, followed by India and Thailand with some 35,000 tonnes each. Other major producer countries are Taiwan, Bangladesh, and Vietnam. In the United States, only a few hundred small farms for *M. rosenbergii* produced about 50 tonnes in 2003. The U.S. is, though, the largest producer of farmed crayfish. In 2003, U.S. farms produced 33,500 tonnes of red swamp crawfish (*Procambarus clarkii*), a crayfish species native to North America.

# Shrimp Farming

The gate of a traditional shrimp farm in Kerala, India which uses the tide to harvest shrimp

Shrimp farming is an aquaculture business that exists in either a marine or freshwater environment, producing shrimp or prawns (crustaceans of the groups Caridea or Dendrobranchiata) for human consumption.

## Marine

Shrimp grow-out pond on a farm in South Korea

Commercial marine shrimp farming began in the 1970s, and production grew steeply, particularly to match the market demands of the United States, Japan, and Western Europe. The total global production of farmed shrimp reached more than 1.6 million tonnes in 2003, representing a value of nearly US$9 billion. About 75% of farmed shrimp is produced in Asia, particularly in China and Thailand. The other 25% is produced mainly in Latin America, where Brazil, Ecuador, and Mexico are the largest producers. The largest exporting nation is Thailand.

Shrimp farming has changed from traditional, small-scale businesses in Southeast Asia into a global industry. Technological advances have led to growing shrimp at ever higher densities, and broodstock is shipped worldwide. Virtually all farmed shrimp are of the family Penaeidae, and just two species – *Litopenaeus vannamei* (Pacific white shrimp) and *Penaeus monodon* (giant tiger prawn) – account for roughly 80% of all farmed shrimp. These industrial monocultures are very susceptible to diseases, which have caused several regional wipe-outs of farm shrimp populations. Increasing ecological problems, repeated disease outbreaks, and pressure and criticism from both NGOs and consumer countries led to changes in the industry in the late 1990s and generally stronger regulation by governments. In 1999, a program aimed at developing and promoting more sustainable farming practices was initiated, including governmental bodies, industry representatives, and environmental organizations.

## Freshwater

Freshwater prawn farming shares many characteristics with, and many of the same problems as, marine shrimp farming. Unique problems are introduced by the developmental lifecycle of the main species (the giant river prawn, *Macrobrachium rosenbergii*). The global annual production of freshwater prawns in 2010 was about 670,000 tons, of which China produced 615,000 tons (92%).

## Animal Welfare

Eyestalk ablation is the removal of one (unilateral) or both (bilateral) eyestalks from a crustacean. It is routinely practiced on female shrimps (or prawns) in almost every marine shrimp maturation or reproduction facility in the world, both research and commercial. The aim of ablation under these circumstances is to stimulate the female shrimp to develop mature ovaries and spawn.

Most captive conditions for shrimp cause inhibitions in females that prevent them from developing mature ovaries. Even in conditions where a given species will develop ovaries and spawn in captivity, use of eyestalk ablation increases total egg production and increases the percentage of females in a given population that participate in reproduction. Once females have been subjected to eyestalk ablation, complete ovarian development often ensues within as little as 3 to 10 days.

## Scallop Aquaculture

Scallop aquaculture is the commercial activity of cultivating (farming) scallops until they reach a marketable size and can be sold as a consumer product. Wild juvenile scallops, or spat, were collected for growing in Japan as early as 1934. The first attempts to fully cultivate scallops in farm environments were not recorded until the 1950s and 1960s. Traditionally, fishing for wild scallops has been the preferred practice, since farming can be expensive. However worldwide declines in wild scallop populations have resulted in the growth of aquaculture. Globally the scallop aquaculture industry is now well established, with a reported annual production totalling over 1,200,000 metric tonnes  from about 12 species. China and Japan account for about 90% of the reported production.

The sea scallop is cultured in the eastern USA

## Cultured Species

There are varying degrees of aquaculture intensity used for different species of scallop. Therefore, cultured species can be divided into operations that are commercially well-established, those in the early commercial stages, those in development or experimental stages and those where potential for commercial farming has been expressed. Some species fall under multiple categories in different world regions.

## Established Commercial Operations

A cultured Weathervane Scallop.

- *Aequipecten opercularis* (United Kingdom, northern France and Spain, Norway)

- *Agropecten irradians* (China)

  Sub species *A. irradians irradians* (eastern USA)

Sub species *A. irradians concentricus* (eastern USA)

- *Agropecten purpuratus* (Chile)

- *Chlamys farreri* (China)

- *Chlamys islandica* (eastern USA)

- *Chlamys nobilis* (Japan, China)

- *Mizuhopecten/Patinopecten yessoensis* (eastern RussiaJapan, China, Western Canada [hybridized with *Patinopecten caurinus*])

- *Pecten fumatus* (Australia)

- *Pecten maximus* (United Kingdom, northern France and Spain, Norway)

- *Placopecten magellanicus* (eastern USA)

## Early Commercial Operations

- *Argopecten ventricosus* (Mexico)

- *Chlamys islandica* (Norway)

- *Crassedoma giganteum* (western North America)

## Developmental or Experimental

Nodipecten nodosus

- *Aequipecten tehuelchus* (Argentina)

- *Agropecten opercularis* (Norway)

- *Euvola ziczac* (Venezuela)

- *Nodipecten nodosus* (Brazil, Venezuela)

- *Patinopecten caurinus* (western North America)

- *Pecten maximus* (China)

## Species with Potential

- *Amusium balloti* (Australia)

- *Chlamys varia* (northern Europe)

- *Chlamys islandica* (northern Europe)

- *Euvola vogdesi* (Mexico)

- *Euvola ziczac* (Brazil)

- *Flexopecten flexosus* (northern Europe)

- *Nodipecten subnodosus* (Mexico)

## Other Species of Note

Attempts at cultivation of *Chlamys hastate* and *Chlamys rubida* in western North America have been halted due to the small size and slow growth of both species. Initial attempts made at cultivation of *Pecten novazelandiae* in New Zealand were hampered by large levels of fouling by mussels and by competition from a largely successful natural fishery.

## Methods of Culture

There are a variety of aquaculture methods that are currently utilized for scallops. The effectiveness of particular methods depends largely on the species of scallop being farmed and the local environment.

## Spat Collection

Collection of wild spat has historically been the most common way obtaining young scallops to seed aquaculture operations. There are a variety of ways in which spat can be collected. Most methods involve a series of mesh spat bags suspended in the water column on a line which is anchored to the seafloor. Spat bags are filled with a suitable cultch (usually filamentous fibers) onto which scallop larvae will settle. Here larvae will undergo metamorphosis into post-larvae (spat). Spat can then be collected and transferred to a farm site for on-growing.

Spat collectors will be set in areas of high scallop productivity where spat numbers are naturally high. However, to establish where the most appropriate areas to collect spat are, trial collectors will often be laid at a variety of different sites. Well-funded farms can potentially set thousands of individual spat bags in collection trials.

## Hatcheries

Scallop hatcheries provide a number of advantages over traditional spat collection for supplying seed to aquaculture operations, most notably in selective breeding and genetic manipulation, as well as providing a regular supply of spat at a low price. While initial attempts to culture scallops in hatcheries were fraught with extremely low spawning and high larval mortality rates, a number of successful techniques have now been developed.

One of the most important aspects of hatchery rearing is obtaining and maintaining broodstock. Broodstock must be conditioned so to stimulate gonad development leading up to spawning and much research has been devoted to identifying the best diets and water quality requirements for broodstock. Once broodstock have been conditioned, spawning can be induced. This is most commonly achieved by varying water temperature, increasing water circulation, or by an injection of serotonin (a neurotransmitter). Following spawning, scallop eggs will develop into the "D" larval (shelled) stage in 2 to 4 days post-fertilization. As larvae, they continue to grow and can be fed a variety of microalgal diets with mixed algal diets being reported as giving higher growth rates than single species diets. Settlement of larvae in hatcheries typically occurs between 35 and 45 days after fertilization of the scallop eggs when larvae are approximately 250 µm in size. Following settlement, the larvae undergo metamorphosis where they rearrange their body form to begin their life as a sea-floor dwelling juvenile scallop. Mortality rates are often highest during metamorphosis as larvae go through a series of behavioral and anatomical changes such as loss of the velum (the larval feeding structure) and development of new filter-feeding mechanisms and gills. Post-settlement spat may be further grown in nursery tanks before being transferred to an on-growing system such as pearl nets.

## Grow Out Stage

There are two recognized systems for the grow out stage of scallops. These are hanging culture and bottom culture. Each has its own benefits and drawbacks in terms of cost, ease of handling and quality of the finished product. Enclosed culture systems still have not yet been fully developed in a commercial context and development continues in this area. Such a system would have large advantages over other forms of culture as tight control of feed and water parameters could be maintained.

## Hanging Culture

Hanging culture relies on either a raft or longline system (with buoys and lines) that floats on the sea surface from which the cultured scallops are suspended, usually on ropes to which they are attached in some manner. Rafts are considerably more expensive than the equivalent distance of longline and are largely restricted to sheltered areas. However, raft systems require much less handling time. Longlines have proved effective for most farms to date and have the added advantage of being able to be completely submerged (with the exception of marker buoys) so to reduce visual pollution. From a raft or longline a variety of culture equipment can be supported. The main advantage of any form of hanging culture is in the exploitation of mid-water algal populations that cannot be fully utilized in other forms of culture.

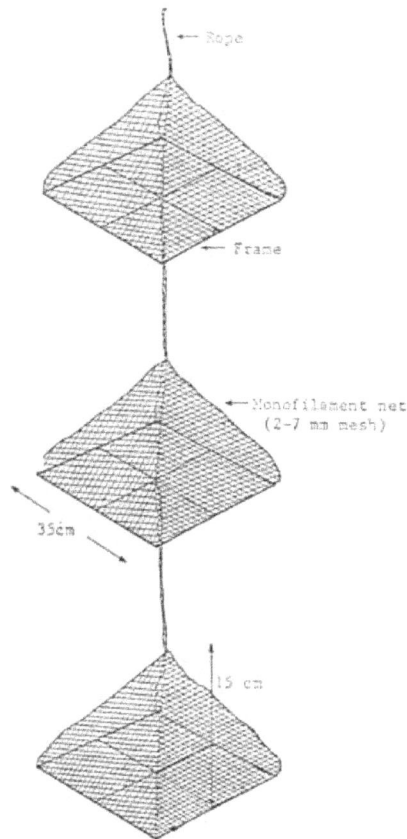

Pearl nets used to grow spat to juveniles.

## Pearl Nets

Once scallop spat have been collected, the most common way of growing them further is in pearl nets (small pyramid shaped nets usually about 350mm across with 2-7mm mesh). Here, they are usually grown to approximately 15mm in high stocking densities. Pearl nets are typically hung ten to a line and have the advantage of being light and collapsible for easy handling. Scallops are usually not grown to larger sizes in pearl nets due to the light construction of the equipment. Once juveniles have reached a desired size they can be transferred to another form of culture.

## Lantern Nets

Lantern nets were first developed in Japan and are the most common method of growing out scallops following removal of juveniles from pearl nets. They allow the scallops to grow to adulthood for harvest due to their larger size and more sturdy construction. Lantern nets are employed in a similar fashion in the mid-water column and can be utilized in relatively high densities. Flow rate of water and algae is adequate and scallops will usually congregate around the edges of the circular net to maximise their food intake.

## Ear Hanging

Ear hanging methods were developed to be a cheaper alternative to lantern nets. Subsequently, research has shown that growth of ear-hung scallops can also be higher than those in lantern nets. Ear hanging involves drilling a hole in the scallop ear (the protruding margin of shell near where the two shells join) and attaching it to a fixed submerged line for growth. Such a process can be relatively labor-intensive as each scallop must be individually handled and drilled (however, many operations now have machines for this process). Furthermore, high mortality rates can result from drilling if scallops are too small, are drilled incorrectly, or spend too much time out of water and become physiologically stressed. This has resulted in research being conducted into the optimal drilling size. This size has been shown to be species specific with small species not having good survival rates. As such, ear hanging is an effective method of growing out larger scallop species. If ear hanging is an appropriate method, scallops can be densely stocked in pairs on lines with as little as 100 mm between pairs. Scallops are maintained in this fashion until harvest. A variety of attachment products are constantly being tested with the best growth so far being obtained with a fastener called a securatie.

## Rope Culture

Rope culture is very similar in principle to ear hanging with the only difference being the method of attachment. In rope culture, instead of scallops being attached by an ear hanging, they are cemented by their flat valve to a hanging rope. This method results in a similar growth and mortality rates as ear hanging but has the advantage of being able to set scallops from a small size. New cementing technologies are being continually developed with the aim of producing quicker setting adhesives to minimize the time scallops spend out of water so to minimize stress.

## Pocket Nets

Pocket netting involves hanging scallops in individual net pockets. Pockets are most often set in groups hanging together. Pocket nets are not used extensively in larger farms due to their cost. However, handling time is low and so can be considered in smaller operations.

## Hog Rigging

Hog rigging involves netting pockets of three or four scallops tied around a central line. This method is quick and cost effective and has been used to a great extent in the European Queen Scallop (*Aequipecten opercularis*) industry. However, its success in larger scallop species has been limited.

## Plastic Trays

Growing scallops in suspended plastic trays such as oyster cages can serve as an alternative to equipment like lantern nets. However, such systems can be expensive and have a rigid structure so cannot be folded and easily stored when not in use. In general, plastic trays are mostly used for temporary storage of scallops and for transportation of spat.

## Bottom Culture

Methods of bottom culture can be used in conjunction with or as an alternative to hanging culture. The main advantage of using methods of bottom culture is in the lower cost of reduced buoyancy needs as equipment is supported by the seabed. However, growing times have been noted as being longer in some cases due to the loss of use of mid-water plankton.

## Plastic Bottom Trays

Plastic trays such as oyster cages can again be utilized in bottom culture techniques. They provide simple and easy to use system. Plastic trays are effective in large numbers but their size is limited by the growth rates of scallops near the centre of cages due to reduced water and food flow rates.

## Wild Ranching

Wild ranching is by far the cheapest of all forms of scallop farming and can be very effective when large areas of seabed can be utilized. However, there can often be problems with predators such as crabs and starfish so areas must be cleared and then fenced to some degree. However, clearing and fencing will not prevent settlement of larvae of predators. Harvesting is usually done by dredge further reducing costs. On smaller farms, however, divers may be employed to harvest.

## Feeding

Scallops are filter feeders that are capable of ingesting living and inert particles suspended in the water column. In culture, scallop diets contain primarily phytoplankton that occur either naturally at a site or are produced artificially in culture. Much research has been conducted into what species of phytoplankton are most effective for inducing growth (and particularly growth of the adductor muscle). Such research has shown that of the species commonly used in bivalve aquaculture, *Isochrysis aff. galbana* (clone T-Iso) and *Chaetoceros neogracile* are the most effective. Recently, with the increase of enclosed farming techniques, a large amount of work has been directed at development of an artificial microalgal substitute that is more cost effective than traditional feeds.

Microalgae cultures may also be manipulated in order to produce algae with a more desirable protein, lipid and carbohydrate profile and much work is being conducted in this area. Furthermore, microalgal species used in scallop culture usually have high levels of vitamins such as vitamin C. The dietary requirements of scallops differ depending on species and life stage. For example, increased protein content of the microalgal diet of broodstock has been shown to reduce time to spawning maturity and increase fecundity. Similar positive results for growth and survival have been observed in larvae fed with high protein diets. However, speculation remains that lipids are also very important to scallop larvae.

## Diseases, Parasites and Phycotoxins

### Diseases

As with any aquaculture species, the incidence of diseases (and parasites) can be amplified by the close proximity of individuals. The occurrence of diseases in scallop culture has been presented as subdued and not well understood; however, the Chinese production of Farrer's scallop (*Chlamys farreri*) was devastated by malacoherpesviridae in the 1990s. Databases are being assembled and industry and research institutions are cooperating for an improved effort against future outbreaks.

## Parasites

A similar situation is seen with parasites as is seen with diseases: at this stage little is known about scallop parasites and few have been identified. As of 2006, no mass deaths caused by parasites have been reported. There are only 17 parasites and commensals that have been described as being associated with scallops.

## Phycotoxins

The occurrence of phycotoxins is generally associated with specific bodies of water and must be considered during establishment of farms as many phycotoxins derived from toxic algae can have detrimental effects on consumers of infected meat. With respect to scallop culture, two categories of toxins have been reported: Paralytic shellfish poisoning (PSP) and amnesic shellfish poisoning (ASP). PSP has been reported for a number of years in *Placopecten magellanicus* in the Northwest Atlantic and so must be considered in culture operations, particularly as *P. magellanicus* is reported as being a slow detoxifyer of the toxin. ASP is a neurotoxin produced by some marine diatoms and has also been reported in scallops from the Northwest Atlantic (Bird & Wright, 1989). Diarrehetic shellfish poisons (DSP) have also been identified as a potential problem, however, they have not yet been reported in scallop culture. DSPs cause gastrointestinal distress in humans and are produced by dinoflagellates that may be ingested by scallops.

## End Product

The finished product: adductor muscle meat of the Giant Scallop, *Pecten maximus*.

Once scallops have been grown, harvested and processed the principal end product is the meat, which usually consists of just the adductor muscle (fresh or frozen). However, it is becoming increasingly popular to sell the muscle with the roe still attached and also to sell whole animals (primarily in North America). Thus, the industry now produces three distinguishably different products.

While the shelf life of a live scallop is limited, the marketing of this product allows scallop farmers to sell smaller animals and so increase cash flow. Top quality scallop muscle can demand a high market price, which fluctuates with production, success of wild scallop fisheries and a number of other global factors.

## Environmental Impacts

Contrary to common perception concerning the negative impacts of many aquaculture practices (particularly finfishes), scallop aquaculture (and indeed other shellfish aquaculture practices) in many parts of the world are considered to be a sustainable practice that can have positive ecosystem effects. This is a result of filter-feeding bivalves removing suspended solids, unwanted nutrients, silt, bacteria and viruses from the water column so to increase water clarity which, in turn, improves pelagic and benthic ecosystems, particularly by promoting growth of vegetation such as seagrasses.

With this considered, such positive impacts are very area specific and one of the main negative environmental impacts scallop culture can create in some other areas is the eutrophication of waters. This has been well observed in Russia where culture of scallops in partially closed bays has resulted in eutrophication and so changes in species composition and structural and functional parameters of pelagic and benthic communities. Monitoring has shown that after farms are disbanded, ecosystems were restored within 5–10 years. This is in line with a large body of data showing bivalve aquaculture activities result in various environmental changes including changes in hydrological regime, ecological communities (including planktonic communities), biochemical composition of waters, biodeposits and invertebrate settlement success. Furthermore, aquaculture farms are a source of visual pollution and so often attract public opposition in highly populated coastal areas.

## Aquaculture of Catfish

Loading U.S. farm-raised catfish

Catfish are easy to farm in warm climates, leading to inexpensive and safe food at local grocers.

Catfish raised in inland tanks or channels are considered safe for the environment, since their waste and disease should be contained and not spread to the wild.

## Asia

In Asia, many catfish species are important as food. Several walking catfish (Clariidae) and shark catfish (Pangasiidae) species are heavily cultured in Africa and Asia. Exports of one particular shark catfish species from Vietnam, *Pangasius bocourti*, has met with pressures from the U.S. catfish industry. In 2003, The United States Congress passed a law preventing the imported fish from being labeled as catfish. As a result, the Vietnamese exporters of this fish now label their products sold in the U.S. as "basa fish".

## United States

Ictalurids are cultivated in North America (especially in the Deep South, with Mississippi being the largest domestic catfish producer). Channel catfish (*Ictalurus punctatus*) supports a $450 million/yr aquaculture industry. The US farm-raised catfish industry began in the early 1960s in Kansas, Oklahoma and Arkansas. Channel catfish quickly became the major catfish grown, as it was hardy and easily spawned in earthen ponds. By the late 60s, the industry moved into the Mississippi Delta as farmers struggled with sagging profits in cotton, rice and soybeans, especially on those farm areas where soils had a very high clay content.

The Mississippi Deltaic Plain includes two active pro-grading deltas: the modern bird-foot [Balize] delta, commonly referred to as the Mississippi Delta, and the Atchafalya delta. In addition, there are degrading deltaic systems, such as the Lafourche and the St. Bernard [ref: World Delta Data Base, Hart and Coleman]. These deltas became the industry home for the catfish industry, as they had the soils, climate and shallow aquifers to provide water for the earthen ponds that grow 360-380 million pounds (160,000 to 170,000 tons) of catfish annually. Catfish are fed a grain-based diet that includes soybean meal. Fish are fed daily through the summer at rates of 1-6% of body weight with the pelleted floating feed. Catfish need about two pounds of feed to produce one pound of live weight. Mississippi is home to 100,000 acres (400 km$^2$) of catfish ponds, the largest of any state. Other states important in growing catfish include Alabama, Arkansas and Louisiana.

## Aquarium

There is a large and growing ornamental fish trade, with hundreds of species of catfish, such as *Corydoras* and armored suckermouth catfish (often called plecos), being a popular component of many aquaria. Other catfish commonly found in the aquarium trade are banjo catfish, talking catfish, and long-whiskered catfish.

# Aquaculture of Tilapia

Tilapia has become the third most important fish in aquaculture after carp and salmon; worldwide production exceeded 1,500,000 metric tons in 2002 and increases annually. Because of their high protein content, large size, rapid growth (6 to 7 months to grow to harvest size), and palatability, a number of tilapiine cichlids—specifically, various species of *Oreochromis*, *Sarotherodon*, and *Tilapia*—are the focus of major aquaculture efforts.

Tilapia

Tilapia fisheries originated in Africa. The accidental and deliberate introductions of tilapia into Asian freshwater lakes have inspired outdoor aquaculture projects in various countries with tropical climates, most notably Honduras, Papua New Guinea, the Philippines and Indonesia. Tilapia farm projects in these countries have the highest potential to be "green" or environmentally friendly. In temperate zone localities, tilapia farmers typically need a costly energy source to maintain a tropical temperature range in their tanks. One relatively sustainable solution involves warming the tank water using waste heat from factories and power stations.

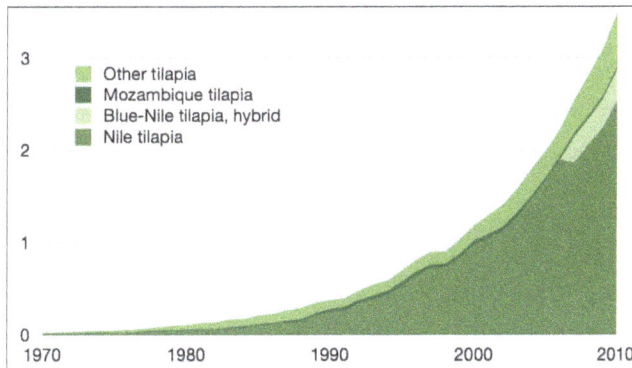

Aquaculture production of tilapia by species
in million tonnes as reported by the FAO, 1950–2009

Tilapiines are among the easiest and most profitable fish to farm due to their omnivorous diet, mode of reproduction (the fry do not pass through a planktonic phase), tolerance of high stocking density, and rapid growth. In some regions the fish can be raised in rice fields at planting time and grow to edible size (12–15 cm, 5–6 in) when the rice is ready for harvest. Unlike salmon, which rely on high-protein feeds based on fish or meat, commercially important tilapiine species eat a vegetable or cereal-based diet.

Tilapia raised in inland tanks or channels are considered safe for the environment, since their waste and disease is contained and not spread to the wild. However, tilapiines have acquired notoriety as being among the most serious invasive species in many subtropical and tropical parts of the world. For example, *Oreochromis aureus*, *O. mossambicus*, *Sarotherodon melanotheron melanotheron*, *Tilapia mariae*, and *T. zilli* have all become established in the southern United States, particularly in Florida and Texas.

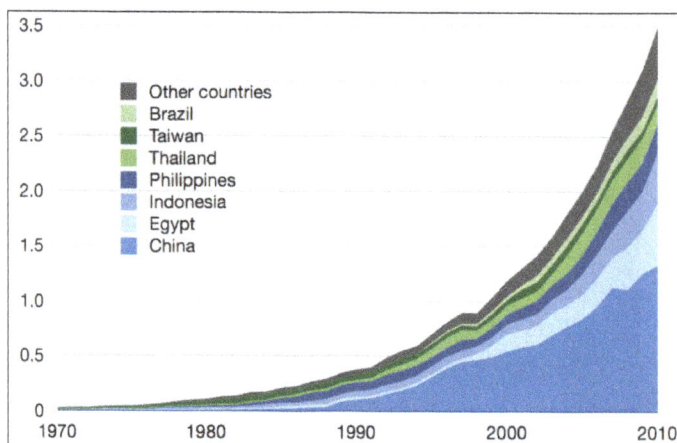

Aquaculture production of tilapia by country
in million tonnes as reported by the FAO, 1950–2009

Commercially grown tilapia are almost exclusively male. Being prolific breeders, female tilapia in the ponds or tanks will result in large populations of small fish. Whole tilapia can be processed into skinless, boneless (PBO) fillets: the yield is from 30% to 37%, depending on fillet size and final trim.

## Nutritional Value

Tilapia from aquaculture contain especially high ratios of omega-6 to omega-3 fatty acids.

## Around the World

Apart from the very few species found in the Levant, such as the Middle Eastern mango tilapia, there are no tilapiine cichlids endemic to Asia. However, species originally from Africa have been widely introduced and have become economically important as food fish in many countries. China, the Philippines, Taiwan, Indonesia and Thailand are the leading suppliers, and these countries altogether produced about 1.1 million metric tonnes of fish in 2001, constituting about 76% of the total aquaculture production of tilapia worldwide.

| Production of farmed tilapia from the top 20 countries in 2010 | | |
|---|---|---|
| Country | Tonnes | Notes |
| China | 1,331,890 | |
| Egypt | 557,049 | |
| Indonesia | 458,752 | In Indonesia, tilapia are known as *ikan nila*. Tilapia were introduced to Indonesia in 1969 from Taiwan. Later, several species also introduced from Thailand (Nila Chitralada),Philippines (Nila GIFT) and Japan (Nila JICA). Tilapia has become popular with local fish farmers because they are easy to farm and grow fast. Major tilapia production areas are in West Java and North Sumatra. In 2006, Badan Pengkajian dan Penerapan Teknologi (Agency for the Assessment and Application of Technology) and Balai Besar Pengembangan Budidaya Air Tawar (Main Center for Freshwater Aquaculture Development – MCFAD), Indonesian government research, development and introduced a new species named "genetically supermale Indonesian tilapia" (GESIT). GESIT fish are genetically engineered to hatch eggs that will produce 98% - 100% male tilapia. Monosex culture (all male) is more productive and will benefit the farmers. Now, around 14 strains of *ikan nila* have been developed by contributions from research institutes including MCFAD. |

| Production of farmed tilapia from the top 20 countries in 2010 | | |
|---|---|---|
| **Country** | **Tonnes** | **Notes** |
| Philippines | 258,839 | In the Philippines, several species of tilapia have been introduced into local waterways and are farmed for food. Tilapia fish pens are a common sight in almost all the major rivers and lakes in the country, including Laguna de Bay, Taal Lake and Lake Buhi. Locally, tilapia are also known as *pla-pla*. Tilapiine cichlids have many culinary uses, including fried, grilled, *sinigang* (a sour soup using tamarind, guava, calamansi or other natural ingredients as a base), *paksiw* (similar to *sinigang*, only it uses vinegar) and many more recipes. On January 11, 2008, the Cagayan Bureau of Fisheries and Aquatic Resources (BFAR) stated that tilapia production grew and Cagayan Valley is now the Philippines' tilapia capital. Production supply grew 37.25% since 2003, with 14,000 metric tons (MT) in 2007. The recent aquaculture congress found the growth of tilapia production was due to government interventions: provision of fast-growing species, accreditation of private hatcheries to ensure supply of quality fingerlings, establishment of demonstration farms, providing free fingerlings to newly constructed fishponds, and the dissemination of tilapia to Nueva Vizcaya (in Diadi town). |
| Thailand | 179,355 | The red hybrid is known in Thai language as *pla taptim* (Thai: ปลาทับทิม), whereas the black and silver striped hybrid is known as *pla nin* (Thai: ปลานิล; lit. "Nile fish"). Both hybrids of tilapia *O. niloticus* are very popular in Thai cuisine. Thailand has its share of fish farms and fish pens devoted to the culture of tilapia species. In March 2007, millions of caged tilapia in the Chao Praya river died as the result of a massive fish kill. The cause for this was determined to be oxygen deprivation on a massive scale, one of the causes for fish kills. |
| Brazil | 155,451 | |
| Viet Nam | 76,000 | |
| Taiwan | 74,888 | In Taiwan, tilapiine cichlids are also known as the "South Pacific crucian carp", and since their introduction, have spread across aquatic environments all over the island. Introduced in 1946, tilapiine cichlids made a considerable economic contribution, not only by providing the Taiwanese people with food, but also by allowing the island's fish farmers to break into key markets, such as Japan and the United States. Indeed, tilapiine cichlids have become an important farmed fish in Taiwan for both export and domestic consumption. The Chinese name for the fish in Taiwan is *wu-kuo* (吳郭), and was created from the surnames of Wu Chen-hui (吳振輝) and Kuo Chi-chang (郭啟彰), who introduced the fish into Taiwan from Singapore. The Taiwan tilapia is a hybrid of *Oreochromis mossambicus* and *O. niloticus niloticus*. In mainland China, it is called *luofei* fish (罗非鱼), named after the origin of this fish: the Nile and Africa (*niLUO* and *FEIzhou* in Chinese respectively). |
| Colombia | 49,893 | |
| Ecuador | 47,733 | |
| Myanmar | 40,583 | |
| Malaysia | 38,886 | |
| Uganda | 31,670 | |
| Bangladesh | 24,823 | |
| Costa Rica | 23,034 | |
| Lao People's Dem. Rep. | 20,580 | |
| Honduras | 16,455 | In Spanish, tilapia are simply known as *tilapia*. Formal tilapia farming is relatively new to Honduras but the commercial export market is expanding rapidly. The first audit of a Honduran tilapia fishery was conducted in 2010 and the facility was found to be compliant with international standards. Honduran aquafarmers are now exporting nearly 20 million pounds of the fish every year, leading tilapia to become viewed as a promising commodity for the developing nation. Joint efforts among community farm training centers, a non-profit Honduran microfinance group, FEHMISSE, and foreign investors are assisting local entrepreneurs as they establish and maintain environmentally sound tilapia farms. |

| Production of farmed tilapia from the top 20 countries in 2010 | | |
|---|---|---|
| **Country** | **Tonnes** | **Notes** |
| Nigeria | 11,989 | |
| Zambia | 10,208 | |
| United States | 9,979 | The geographic range for tilapia culture is limited by their temperature-sensitivity. For optimal growth, the ideal water temperature range is 82 to 86 °F (28 to 30 °C), and growth is reduced greatly below 68 °F (20 °C). Death occurs below 50 °F (10 °C). Therefore, only the southernmost states are suitable for tilapia production. In the southern region, tilapia can be held in cages from five to 12 months per year, depending on location. About 1.5 million tons of tilapia were consumed in the US in 2005, with 2.5 million tons projected by 2010. |
| Other countries | 79,335 | |
| TOTAL PRODUCTION | 3,497,391 | |

## Other Countries

## India

The FAO has not recorded any production of farmed tilapia by India. Rajiv Gandhi Centre for Aquaculture (RGCA), the R&D arm of Marine Products Export Development Authority, has established a facility in Vijayawada to produce mono-sex tilapia in two strains. This project involves the establishment of a satellite nucleus for the GIFT strain of tilapia in India, the design and conduct of a genetic improvement program for this strain, the development of dissemination strategies, and the enhancement of local capacity in the areas of selective breeding and genetics. The development and dissemination of a high yielding tilapia strain possessing desirable production characteristics is expected to bring about notable economic benefits for the country. Farming of Tilapia is not permitted in the country on commercial basis. The Rajiv Gandhi Center for Aquaculture (RGCA) has expressed interest in obtaining the Genetically Improved Farmed Tilapia (GIFT strain) for aquaculture development in the country. The GIFT tilapia strain, selectively bred in Malaysia and the Philippines, has achieved an improvement of more than 10 per cent per generation in growth rate and has been widely distributed to several Asian countries and to Latin America (Brazil). However, rather than passively importing the improved genetic stock, the Center is interested in running a formal breeding program (fully pedigreed population) similar to the one that has been carried out for the GIFT strain in Malaysia.

The aim is to produce fast-growing high yielding tilapia strains adapted to a wide range of local farming environments that can be grown at as low a cost as possible.

The project involves several steps. The first is the establishment of a new nucleus of the GIFT strain at the RGCA and the design of a formal breeding program to further improve its genetic performance within the local environment. This will involve enhancing the capacity of local personnel in selective breeding, genetic improvement, statistical analysis and hatchery management through specialized training courses.

Once a high performing tilapia strain (or strains) has been developed, the establishment of satellite hatcheries will increase the availability and decrease the costs of seed stock. These public and private hatcheries will act as multipliers for the superior genetics developed at RGCA and the sites for dissemination of quality broodstock to fish farmers.

Although the ultimate target groups of this project are fish farmers and small householders, a wider range of beneficiaries is expected, including commercial producers, scientists and the end consumers. The RGCA will gain experience and knowledge on the development of genetic improvement programs for economically important traits and other aspects of modern quantitative genetics. This experience and the development of a standard selective breeding protocol will allow for genetic improvement programs for other aquaculture species that are commonly cultured in India. Hatchery managers, producers and farmers will also improve their capacity to implement on-farm selective breeding programs.

In the longer term the project is also expected to contribute to the development of a complete chain of production. This will require initial capital support for farmers, identification of alternative cheap plant-based feed, and diagnosis of diseases in hatcheries, as well as strategies for early growth management. Improvement in harvest technologies, including storage of product and transport facilities, is likely to improve as a consequence of this project.

## Malawi

In 2010 Malawi produced 2,997 tonnes of farmed tilapis. A variety of tilapia, Oreochromis lidole is one of the most popular fish in Malawi. It is locally known as 'chambo' in Malawi. It is endemic to bodies of water in Malawi like Lake Malawi, Lake Malombe and the Shire River. Due to over fishing, the fish however is now on the threatened species list. Malawi has its fish farms that are dedicated to farming this fish.

## Philippines

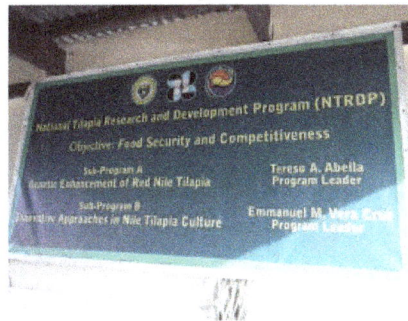

National Tilapia Research and Development Program (NTRDP) – Red Nile Tilapia
Aquaculture of tilapia  Nile tilapia Oreochromis niloticus

Red nile tilapia under the experiment (CLSU), Philippines.)

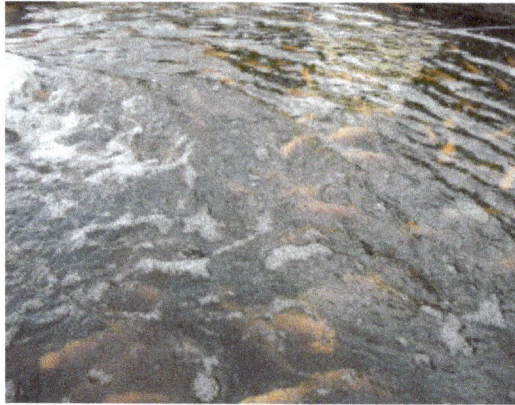

Red nile tilapia for breeding

The fingerlings in breeding tubes

Water sprinklers for aeration of the fingerlings

# Aquaculture of Sea Cucumbers

Sea cucumber stocks have been overexploited in the wild, resulting in incentives to grow them by aquaculture. Aquaculture means the sea cucumbers are farmed in contained areas where they can be cultured in a controlled manner. In China, sea cucumbers are cultured, along with prawns and some fish species, in integrated multi-trophic systems. In these systems, the sea cucumbers feed on the waste and feces from the other species. In this manner, what would otherwise be polluting byproducts from the culture of the other species become a valuable resource that is turned into a marketable product.

Sea cucumbers are usually scavengers which feed on the debris on the sea floor

## History

The Chinese and Japanese were the first to develop successful hatchery technology for *Apostichopus japonicus*, prized for its high meat content and success in commercial hatcheries. A second species, *Holothuria scabra*, was cultured for the first time using these techniques in India in 1988. In recent years Australia, Indonesia, New Caledonia, Maldives, Solomon Islands and Vietnam have also successfully cultured *H. scabra* using the same technology, which has since been expanded to other species.

## Broodstock

Sandfish hatchery (Alaminos, Pangasinan).

Sea cucumbers to be used as broodstock are either collected from the wild or are taken from commercial harvests. Only the largest and healthiest individuals are used for broodstock, as the success of a hatchery relies on the healthy condition of brood individuals. These individuals are kept in tanks with at least 6 inches of sand to allow burrowing behaviour. Water is changed every day and sand is changed every fortnight. Sea cucumbers are fed with a paste made from freshly collected algae added to the tanks once a week to settle on sand where they feed. If water conditions are not right and if proper food is not provided sea cucumbers will eviscerate or re-absorb their gonads rendering them unfit for spawning.

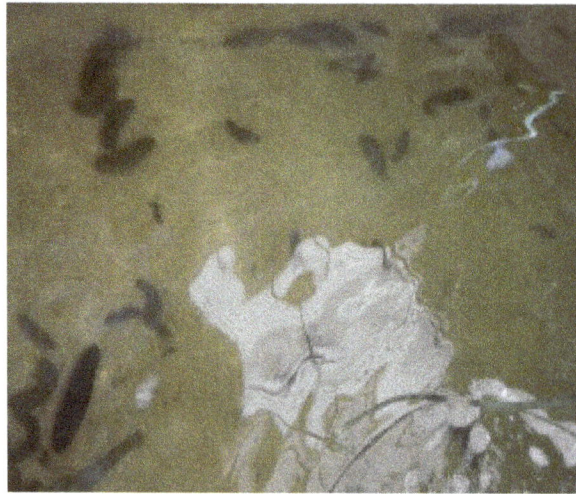

The Philippine "Balatan" or Sea cucumber breeding/harvesting.

## Spawning

Temperature shock involves cooling and heating of seawater by 3–5°C until spawning is induced. This is achieved by first reducing the temperature of the water by 3–5°C. The sea cucumbers are left for five minutes before they are exposed to 'normal' (depending on species and climate) temperature seawater, where the small rise in water temperature is sufficient to induce spawning. Males tend to spawn first which then induces females to release their eggs.

Spawning stimulation can also be achieved through lightly drying the broodstock followed by exposure to a powerful jet of seawater. Sea cucumbers are dried for 30 minutes in the shade and then are exposed to a powerful jet of seawater for 30 minutes. Usually 60–90 minutes later males will release their sperm, and 30 minutes after that females will swell and release eggs in rapid intermittent jets.

Though many species of sea cucumbers can be induced to spawn using both of these methods, temperature shock is usually considered to be the preferred method. Often spawn obtained from drying and wetting with a jet of water does not produce viable gametes. Spawning induction and successful fertilization has only been achieved in some species of sea cucumbers and the likelihood that a method will work or not is highly dependent on the species.

## Larvae

The first month after hatching is particularly crucial and mortality during the larval phases is particularly high. Larval survival drops to 30 – 34% after the first 20 days of hatching and larval development. Larvae usually hatch 48 hours after fertilisation and spend their first 17 days as feeding larvae or auricularia. During this phase they are fed on a mix of planktonic microalgae (*Rhodomonas salina*, *Chaetoceros calcitrans*, *C. mulleri*, *Isochrysis galbana* and *Pavlova lutheri* are most commonly used). The proportions and overall quantity of microalgal feed species varies with larval stage, and the quantity is gradually increased as larvae grow until they metamorphose into the doliolaria or non-feeding phase (around day 17.) Individuals in this phase of their development are put into a tank with settlement cues. These may include food items such as seagrass extract, seaweed extract, Algamac2000, Algamac Protein Plus, dead algae, benthic diatoms (*Nitzchia sp.* and *Navicula sp.*) and spirulina.

Around day 19 of development the larvae transform into their pentacula phase and settle. Plates or polythene sheets are provided as substrate for larvae to settle on and to feed off. Benthic diatoms *Nitzchia* sp. and *Navicula* sp. are most effective as settlement cues.

## Nursery

Juveniles are sometimes transferred to a sand-based feeding substrate in nursery tanks when they reach 10 mm; however, survival of juveniles is better if they are allowed to grow to 20 mm before transferral to sand. Juveniles are grown for a few months until they reach 5–7 cm when they are moved out to sea ranches or into ponds.

## Grow Out

Sea ranching is carried out in sheltered bays with seagrass in areas with few predators. The sea cucumbers can be kept in pens in shallow water made of fine wire mesh or bamboo, and in deeper water they are raised in cages made from fine woven mesh or in tub enclosures on the seafloor. They can also be kept and grown in ponds with appropriate water exchange and movement. Individual growth is density-dependent and is stunted at high densities. Monitoring water quality and growth characteristics are essential to survival during this phase. Sea cucumbers are ready to harvest after 12 months of grow out.

## Asexual Methods

Two sea cucumber species *Thelenota ananas* (prickly redfish) and *Stichopus chloronotus* (green-fish) have been found capable of asexual propagation through transverse fission, the process whereby an organism is cut in half and completely regenerates the missing half. Rubber bands are placed around the middle of the sea cucumbers which induces them to undergo fission within 1–2 weeks. After separating, the posterior half regrows a complete anterior half, and vice versa. This happens within 3–7 months, producing two new fully-grown individuals from one. Survival from this process by these species was found to be 80% or greater. Though this technique is not suitable for all sea cucumber species, it may provide a cheaper and faster alternative method of obtaining prickly redfish and greenfish for aquaculture.

The *prickly redfish* (left) and *greenfish* (right)
can be asexually propagated

## Polyculture

Sea cucumbers are currently cultured in polyculture with prawns and some fish species. Their presence in the bottoms of the pens or nets, where they feed on debris composed of feces, excess food, algae, and other particulate organic matter, significantly reduces fouling of water and equipment. China currently produces around 90,000 tons of sea cucumbers using these practices and enhanced growth of sea cucumber juveniles has been reported when they are grown at the bottom of prawn farms. Farming sea cucumbers with the fouling debris of other aquaculture species helps to mitigate the impacts of marine farms' effluents and turns these wastes into a marketable product.

## Aquaculture of Sea Sponges

Sea sponge aquaculture is the process of farming sea sponges under controlled conditions. It has been conducted in the world's oceans for centuries using a number of aquaculture techniques. There are many factors such as light, salinity, pH, dissolved oxygen and the accumulation of waste products that influence the growth rate of sponges. The benefits of sea sponge aquaculture are realised as a result of its ease of establishment, minimum infrastructure requirements and the potential to be used as a source of income for populations living in developing countries. Sea sponges are produced on a commercial scale to be used as bath sponges or to extract biologically active compounds which are found in certain sponge species. Techniques such as the rope and mesh bag method are used to culture sponges independently or within an integrated multi-trophic aquaculture system setting. One of the only true sustainable sea sponges cultivated in the world occur in the region of Micronesia, with a number of growing and production methods used to ensure and maintain the continued sustainability of these farmed species.

## History

More than 8000 species of sea sponges live in oceanic and freshwater habitats. Sponge fishing historically has been an important and lucrative industry, with yearly catches from years 1913 to 1938 regularly exceeding 181 tonnes and generating over 1 million U.S. dollars. However, this demand for sea sponges has seen catch rates peak and in 2003 the demand for bath sponges was 2,127 tonnes, with global production from harvesting only meeting a quarter of that amount.

Early aquaculture research into optimising techniques for sea sponge aquaculture used a number of farming methods. However, commercial sponge farming was met with severe resistance and interference from sponge fisherman, who believed that their continued income was under threat. The opposition by commercial sponge farmers resulted in a low market penetration and poor consumer adoption of aquacultured sponge products.

## Benefits

The benefits of commercial sponge aquaculture are apparent for those living in developing countries. In these countries, sponge aquaculture is both an easy and profitable business, which benefits the local community and environment through minimising both harvesting pressure on wild stocks and environmental damage.

## Simple

Growing sponges is a relatively simple process and requires little specialist knowledge. Furthermore, the ease of sponge aquaculture means that the whole family can be involved in the production process. This results in a profitable family business which conforms to traditional discourses of "family farms", increasing the likelihood of sea sponge aquaculture adoption. In addition, it is common for sea sponge farms to be located close to family homes allowing for continual access, monitoring, modification and work to be completed on the farm.

## Income Generation

Sea sponge aquaculture also provides families with a continuous source of income year-round, which can be undertaken as a full-time commitment, or as a part-time job to supplement an existing income.

## Uses

## Bath Sponges

The last two decades have seen a renewed interest in the potential for sponge aquaculture to contribute to supplying the growing global demand for bath sponges. Bath sponges are the most common use of aquacultured sea sponge today. Bath sponges can be defined as any sponge species possessing only spongin fibers – which are springy fibres made from collagen protein.

Commercial uses for bath sponges range from cosmetic, bath, or industrial purposes, with the quality of the sponge based on analysing the quality of the sponge skeleton, with those possessing soft, durable and elastic fibres demanding the highest price.

## Bioactive Uses

The presence of secondary metabolites produced by symbiotic microorganisms within the sponge, enhances its growth and survival. Thousands of sponge derived secondary metabolites have been successfully isolated from sponges, with many metabolites having potential medicinal properties, such as cytotoxicity, anti-inflammatory and anti-viral activity. Therefore, they have significant potential within the pharmaceutical industry as a means of generating novel drugs. These secondary metabolites, however, are often only present in trace amounts, with the only methods to use these metabolites as therapeutics depending on the scale up of the compounds via chemical synthesis or aquaculture.

## Menstrual Sponges

While it is still something of a niche market, a few companies have begun to produce and market small sponges as reusable feminine hygiene products. They are marketed under the brand names Sea Pearls in the United States and Jam Sponge in the United Kingdom. The sponges are inserted into the vagina in much the same way a tampon is, but when full are removed, cleaned, and reused, rather than discarded. The advantages of a reusable tampon alternative include cost-effectiveness and waste reduction. (Since sponges are biodegradable, even when a menstrual sponge's absorbent life is over it can be composted.) Some women are also concerned by the health risks associated with traditional tampons and feel it is healthier to use a natural material. While no known cases of toxic shock syndrome have been associated with the use of menstrual sponges, sponges are known to contain sand, grit and bacteria, and thus the possibility of toxic shock syndrome must be considered. Sponges have a greater capacity for absorbing menstrual flow than most tampons; though they should still be changed at least every eight hours.

## Factors that Affect the Growth of Sponges

### Salinity, pH, Temperature and Light

Sea sponges should be cultured at a salinity of 35ppt (salinity of seawater). Hypersalinity (high salt concentrations) in the immediate environment surrounding a sponge will dehydrate sponge cells, whereas a hyposaline (low salt concentration) environment dilutes the intracellular environment of the sponge. The pH of water must match that of seawater (pH 7.8–8.4) in order for sponge production to be maximized. Sponges are sensitive to temperature, and extreme fluctuations in ambient temperature can negatively affect the health of sea sponges. High temperatures lead to crashes in sponge cultures. Symbiotic bacteria that normally inhabit sea sponges start reproducing at an unsustainable rate when ambient temperature of the water increases by a few degrees. These bacteria then attack and destroy the sponge cells and tissue. It has been suggested that sponges should be cultured at water temperatures slightly below the ambient water temperature in the region the sponge has been originally isolated from.

Photosynthetic endosymbionts inhabit many tropical sponges, and these require light to survive.

Certain sponges as a result depend on light availability and intensity to achieve their nutritional needs. In some species however, light may lead to growth inhibition as they are sensitive to ultra-violet radiation. Other than when the sponge has associated photosynthetic bacteria, optimal sea sponge growth occurs in dark conditions.

## Dissolved Oxygen

Dissolved oxygen is absorbed through the aquiferous system. Oxygen in sea sponges is consumed at rates which range from 0.2–0.25 µmol $O_2h^{-1}$/cm$^3$ of sponge volume. Demosponges maintained under laboratory conditions can also tolerate hypoxic conditions, for brief periods, which could reflect their adaptability to dissolved oxygen.

## Waste Removal

In closed culture systems some species of sponge may produce bioactive and cytotoxic metabolites which may rapidly build up and inhibit further sponge growth. However, biofilters are likely to be ineffective at removing secondary metabolites expelled from the sponge. Adsorption methods where biomolecules adhere to an adsorbate are likely to be an effective way of removing these compounds.

## Diseases

Bath sponge disease outbreaks are often severe, having the potential to destroy both wild and aquacultured sponge populations. The underlying factors that result in disease outbreaks may be due to causative agents such as viruses, fungi, cyanobacteria and bacterial strains.

## Site Selection

When choosing a sea sponge aquaculture location, factors that promote growth and survival of the cultured sponge species must be considered. Sponges rely greatly on a passive flow of water to provide food, such as bacteria and microalgae, thus good water flow increases growth and quality of sponges. Higher than normal water flow rates could potentially damage farmed sponges. An ideal location for a sea sponge farm would be in an area that is sheltered, but which receives ample water flow and food availability to optimise sponge growth.

## Methods of Cultivation

### The Use of Explants

Sponge aquaculture for spongin or metabolite production capitalises on the high regenerative abilities of the totipotent sponge cells by using explants (cut pieces of a parent sponge, which will then regrow into a full sponge) as a means of culturing sponges. Sponges have indeterminate growth, with maximum growth determined through environmental constraints rather than genetics. During the initial establishment of a farm, sponge explants will be chosen by their phenotypic characteristics of fast growth and high quality spongin or metabolites.

## Integrated Multi-trophic Aquaculture

Intensive marine aquaculture in the last decade has increased considerably and resulted in considerable adverse environmental impacts. Large discharge volumes of organic matter from uneaten feed and excretory waste from aquacultured species has resulted in high levels of nutrients within coastal waters. Large quantities of nitrogen (~ 75%) excreted from bivalves, salmon and shrimp, enter into the coastal environment, with the potential to develop algal blooms, and reduce dissolved oxygen in the water.

An integrated aquaculture system consists of a number of species at different trophic levels of the food chain. Thus waste generating (fed organisms) such as fish and shrimp are coupled with extractive organisms such as abalone, sponges or sea urchins, as a mechanism of removing excess nutrient matter from the water column. Sea sponges have a distinct advantage as an extractive organism in an integrated multi-trophic aquaculture system, as they have the potential of acting as a bioremediator to remove both pathogenic bacteria and organic matter. The sponge *Hymeniacidon perlevis* has exhibited an excellent ability to remove total organic carbon (TOC) from seawater under integrated aquaculture conditions, and could be a potentially useful bioremediation tool for aquaculture systems in regions where water pollution is high. Furthermore, the organic enrichment originating from fish farmed in the vicinity may stimulate sponge growth, resulting in more efficient sea sponge aquaculture.

## Bath Sponge Aquaculture

Many commercial sea sponge farms situate their aquaculture sites in deeper waters (>5 m), to maximise the number of sponge explants that can be grown and increase productivity. Two main methods of bath sponge aquaculture have been trialled with sponges either being grown on a rope or inside a mesh bag.

## Rope Method

Survival for sponges farmed on ropes is generally lower as unrecoverable damage occurs to the explant when 'threading' onto the rope takes place. Furthermore, sponges cultured on the rope have the potential to be torn off the rope during storms as water flow increases significantly, or grow away from the rope and form an unmarketable, low value, characteristic doughnut shaped sponge. Differences in sponge growth and health do occur within species characterised by variations in regenerative ability, susceptibility to infection after cutting, hardiness and growth potential.

## Mesh Bag Method

Lower levels of damage in some species of sponges cultured via mesh bags can lead to higher levels of survival. Growth rates may be decreased as mesh strands on the bags may decrease water flow, limiting food availability. The accumulation of biofouling agents such as bryozoans, ascidians and algae on the mesh may further limit water flow. Thin mesh strands with large gaps and a well-positioned site may act as a means to mitigate against biofouling and reduced flow rates.

## Combination of Methods

By combining both rope and mesh bag approaches to bath sponge aquaculture in a "nursery period", increases may occur in quality and production. In the nursery period method, sponges are initially cultivated in mesh bags until the explants have healed and regenerated to efficiently filter water. The regenerated explants are transferred onto rope to promote optimal growth till harvesting. This strategy is labour-intensive and costly, with growth rates and survival found to be no greater than when farming occurs solely via the mesh bag method.

A more economically viable method for cultivating bath sponges would be transferring sponges to larger mesh bags as sponge growth occurs to enable adequate water flow and nutrient sequestration.

## Bath Sponge Aquaculture Production in Micronesia

Bath sponges are currently being produced using the sponge *Coscinoderma matthewsi* with production of about 12,000 sponges, sold locally to residents and tourists in Pohnpei, Federated States of Micronesia. These sponges are one of the only true sustainably farmed sea sponges in the world. The sponges are farmed via the rope method, with low investment costs of a few thousand dollars for farming and maintenance equipment, producing 100% natural sponges with no harsh chemicals added during processing.

Aquaculture production of *C. matthewsi* sponges was undertaken by the Marine and Environmental Research Institute of Pohnpei (MERIP), to try and generate a sustainable income for local community residents with few options to earn money. The sponges take approximately two years to reach harvestable size, with free divers routinely removing seaweed and biofouling agents by hand. These sponges are processed through natural processes, where they are left to air dry and then placed in baskets and returned to the lagoon where they were grown. This process removes all the organic matter within the sponge leaving behind the final bath sponge product. Further processing occurs by softening the sponge, but no bleaches, acids or colorants are used.

## Bioactive Sponge Aquaculture

Research into farming sea sponges for bioactive metabolites occurs in the Mediterranean, Indo-Pacific, and South Pacific regions. The main goals are to optimise bioactive production methods, aquaculture processes and environmental conditions to maximise their production.

## New Methods

In the aquaculture for bioactives, the final explant shape is not of concern, allowing for additional production methods to be utilized. New methods of bioactive cultivation include the "mesh array method" which utilises the water column to vertically hang a mesh tube with single explants held in alternating pockets.

The number of sponges required to aquaculture bioactives is reduced as sponge secondary metabolites can be repetitively harvested for many years, decreasing the costs and infrastructure required. The few sponges selected for metabolite production would have high production rates for the target metabolite to optimise production and profits.

## Factors Affecting Secondary Metabolite Production

A number of factors affect sponge metabolite production, with metabolite concentration varying greatly between neighbouring explants. Localised differences in light intensity and bio-fouling are physical and biological factors that have been found to significantly affect metabolite biosynthesis in sponges. Changes in environmental factors may alter microbial populations and subsequently affect metabolite biosynthesis.

Understanding the environmental factors that affect metabolite biosynthesis or the ecological role of the metabolite, can be used as a competitive advantage to maximise metabolite production and total yield. For example, if the ecological role of the secondary target metabolite was to deter predators, mimicking predation via wounding the sponge before harvesting may be an efficient technique to maximise metabolite production.

Some sponges producing metabolites grow extremely quickly, suggesting that farming sponges may be a viable alternative to producing bioactives that at present cannot be chemically synthesised. Although sponge farming for bioactives is more lucrative owing to its higher value-adding properties, there are several challenges that are not present when aquaculturing bath sponges, such as the high costs associated with metabolite extraction and purification.

# Aquaculture of Giant Kelp

Giant kelp

Giant kelp, *Macrocystis pyrifera*, has been utilized for many years as a food source; it contains many compounds such as iodine, potassium, other minerals vitamins and carbohydrates and thus has also been used as a dietary supplement. In the beginning of the 20th century California kelp beds were harvested as a source for potash. With commercial interest increasing significantly during the 1970s and the 1980s this was primarily due to the production of alginates, and also for biomass production for animal feed due to the energy crisis during that period. However commercial production for *M.pyrifera* never became a reality. With the end of the energy crisis and the decline in prices of alginates, the research into farming *Macrocystis* also declined.

## Present

The demand for *M.pyrifera* is increasing due to the newfound uses of these plants such as fertilizers, cultivation for bioremediation purposes, abalone and sea urchin feed. There is current research going into utilizing *M.pyrifera* as feed for other aquaculture species such as shrimps. The supply of *M.pyrifera* for alginate production relied heavily on restoration and management of natural beds during the early 1990s. Other functions such as substrate stabilizing ability of this species was also recognized in California, called the "Kelp bed project" where the adult plants of 3-6m in length were transplanted to increase the stability of the harbor and promote diversity; this was done in efforts to restore the natural environment after extensive harvesting.

With the global demand for aquatic plants increasing there have been great advances with the technology and methods of cultivating them. China and Chile have taken it on a broader scale; these two countries are currently the largest producers of aquatic plants each producing over 300,000 tones each in 2007. Data on how much of this total can be attributed to actual *M.pyrifera* harvesting is sketchy because as opposed to animals where the details on individual species harvested is kept on record, aquatic plants however are usually lumped in to a single category. Both these countries culture a variety of species, in Chile 50% of the production involves several species of Phaeophytes and the other 50% results from harvesting Rhodophytes. China produces a larger variety of seaweeds which also includes chlorophytes. There are also experiments undergoing in Chile to produce hybrids between this species and *M.integrifolia* in efforts to produce a super cultivar.

## Culturing Methods

The most common method of cultivating *M.pyrifera* was developed in China in the 1950s called the long line cultivation system. Where the sporelings are produced in a cooled water greenhouse and then later planted out in the ocean attached to long lines. The depth to which they are grown is varied in different countries. Since this species has an alternation of generations in its life cycle, which involves a large sporophyte and a microscopic gametophyte. The sporophyte is what is harvested as seaweed. The mature sporophyte form the reproductive organs called sori, these are found on the underside of the leaves and produce the motile zoospores that germinate into the gametophyte. To induce sporalation, the selected plants are dried from a couple to up to twelve hours and placed in a seeding container filled with cool seawater of about 9-10 °C; salinity of 30% and a pH of 7.8-7.9. Photoperiod is also controlled for during the sporolation and the growth phases. A synthetic twine of about 2 – 6mm in diameter is placed on the bottom of the same container after sporalation and the released zoospores attach themselves to the twine and begin to geminate into male and female gametophytes. Upon maturity these gametophytes release sperm and egg cells that fuse in the water column and attach themselves to the same substrate as the gametophytes (i.e. the synthetic twine). These plants are then reared up into young sporophyte plants for up to 60 days.

These strings are either wrapped around or are cut up into small pieces and attached to a larger diameter cultivation rope. The cultivation ropes vary but are approximately 60m with floating buoys attached. The depths at which they are grown in the water column vary for some of the countries. In China, *M.pyrifera* is cultivated on the surface with floating buoys attached every 2-3m and the ends of the rope attached to a wooden peg anchored to the substrate, individual ropes are usually hung at 50 cm intervals to each other. In Chile however *M.pyrifera* is grown at a depth of 2m using buoys to keep the plants at a constant depths. These are then left alone to grow until ready to harvest. There are several problems with this method of cultivation as there are difficulties that lay in the management form the transition in the juvenile stages; from spore the gametophyte and embryonic sporophyte which are all done on a land based facility with careful control of water flow, temperature, nutrients, and light. The Japanese use a force cultivation method where a 2-year growth rate is achieved within a single growing season by controlling for the above requirements.

In China a project for offshore or deep water cultivation was also looked at where various farm structures were designed to facilitate the growth of *M.pyrifera*; nutrients from the deep waters were pumped up into the growing kelps. The greatest benefit for this approach was that the algae were released from size constrains that are found in shallow waters. Issues with operational and farm designs plagued the project for deep water cultivation, and prevented further cultivation in this manner.

## Harvesting

The duration of the cultivation is varied upon the region and intensity of the farming, this species is usually harvested after two growth seasons (2 years). For *M.pyrifera* that is artificially cultivated on ropes, they are harvested by a pulley system that is attached on boats that pull the individual lines on the vessels for cleaning. Other countries such as the United States of America (USA) which rely primarily on naturally grown *M.pyrifera* use boats to harvest the surface canopy, the surface canopy is harvested several times per year this is possible due to the fast growth of this species and the vegetative and reproductive parts are left undamaged.

## Control

In the UK, legislation defines giant kelp as a plant which should not be allowed to grow in the wild and these kelp are mechanically removed.

## References

- Mumford, T.F. and Miura, A. 4.Porphyra as food: cultivation and economics. in Lembi, C.A. and Waaland, J.R. 1988. Algae and Human Affairs. Cambridge University Press, Cambridge. ISBN 0-521-32115-8

- Abbott, I A & G J Hollenberg. (1976) Marine Algae of California. California: Stanford University Press. ISBN 0-8047-0867-3

- Hoek, C van den; D G Mann & H M Jahns. (1995) Algae: An Introduction to Phycology. Cambridge: Cambridge University Press. ISBN 0-521-30419-9

- Lobban, C S & P J Harrison. (1994) Seaweed Ecology and Physiology. Cambridge: Cambridge University Press. ISBN 0-521-40334-0

- Mondragon, Jennifer & Jeff Mondragon. (2003) Seaweeds of the Pacific Coast. Monterey, California: Sea Challengers. ISBN 0-930118-29-4

- "Ultrasound, a new separation technique to harvest microalgae - Springer". Springerlink.com. 2003-03-01. Retrieved 2013-08-29.

- Rumble, JM; Hebert, KP; Siddon, CE (2012). "Estimating Geoduck Harvest Rate and Show Factors in Southeast Alaska". In: Steller D, Lobel L, eds. Diving for Science 2012. Proceedings of the American Academy of Underwater Sciences 31st Symposium. Retrieved 2013-09-29.

- Starckx, Senne (31 October 2012) A place in the sun - Algae is the crop of the future, according to researchers in Geel Flanders Today, Retrieved 8 December 2012

- Straus, K.M.; Crosson, L.M.; Vadopalas, B. "Effects of Geoduck Aquaculture on the Environment: A Synthesis of Current Knowledge" (PDF). Washington Sea Grant, University of Washington. Retrieved 25 August 2012.

- Marshall, Robert. "Broodstock Conditioning and Larval Rearing of the Geoduck Clam (Panopea generosa Gould, 1850)" (PDF). (PhD dissertation, The University of British Columbia). Retrieved 22 August 2012.

- "Habitat Conservation Plan for the Washington State Department of Natural Resources' Wild Geoduck Fishery" (PDF). Washington State Department of Natural Resources. Retrieved 20 August 2012.

- Dunagan, Christopher (21 February 2011). "State requires shoreline programs to incorporate geoduck farming standards". Kitsap Sun. Retrieved 25 August 2012.

- Hand, C.; Marcus, K. "Potential Impacts of Subtidal Geoduck Aquaculture on the Conservation of Wild Geoduck Populations and the Harvestable TAC in British Columbia" (PDF). Fisheries and Oceans Canada. Retrieved 20 August 2012.

- McLeary, Anthea (1 September 2012). "Ugly ducklings could grow to $1b aquaculture industry". The National Business Review. Retrieved 1 September 2012.

- Bower, S.M. "Geoduck clam (Panopea abrupta): Anatomy, Histology, Development, Pathology, Parasites and Symbionts: Pathology, Parasites and Symbionts Overview". Fisheries and Oceans Canada. Retrieved 20 August 2012.

- "Geoduck aquaculture: Estimated cost and returns for sub-tidal culture in B.C." (PDF). Aquaculture Factsheet, June 2005, No. 05-01. British Columbia Ministry of Agriculture and Lands. Retrieved 1 September 2012.

# 4

# Aquatic Ecosystem: An Overview

Since no species lives on its own, it is necessary to understand its natural habitat, its prey and its predators. Aquaculture requires the natural interaction of aquatic elements with the species that is being cultivated. The study of aquatic ecosystems provides knowledge that proves invaluable for the growth and sustenance of the farmed species.

## Aquatic Ecosystem

An estuary mouth and coastal waters, part of an aquatic ecosystem

An aquatic ecosystem is an ecosystem in a body of water. Communities of organisms that are dependent on each other and on their environment live in aquatic ecosystems. The two main types of aquatic ecosystems are marine ecosystems and freshwater ecosystems.

## Types

### Marine

Marine ecosystems cover approximately 71% of the Earth's surface and contain approximately 97% of the planet's water. They generate 32% of the world's net primary production. They are distinguished from freshwater ecosystems by the presence of dissolved compounds, especially salts, in the water. Approximately 85% of the dissolved materials in seawater are sodium and chlorine. Seawater has an average salinity of 35 parts per thousand (ppt) of water. Actual salinity varies among different marine ecosystems.

Marine ecosystems can be divided into many zones depending upon water depth and shoreline features. The oceanic zone is the vast open part of the ocean where animals such as whales, sharks, and tuna live. The benthic zone consists of substrates below water where many invertebrates live. The intertidal zone is the area between high and low tides; in this figure it is termed the littoral zone. Other near-shore (neritic) zones can include estuaries, salt marshes, coral reefs, lagoons and mangrove swamps. In the deep water, hydrothermal vents may occur where chemosynthetic sulfur bacteria form the base of the food web.

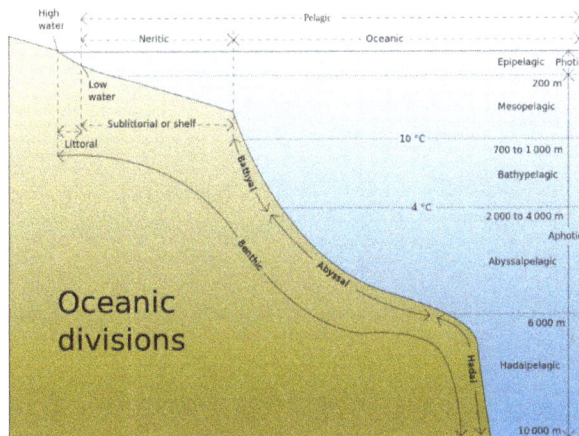

A classification of marine habitats.

Classes of organisms found in marine ecosystems include brown algae, dinoflagellates, corals, cephalopods, echinoderms, and sharks. Fishes caught in marine ecosystems are the biggest source of commercial foods obtained from wild populations.

Environmental problems concerning marine ecosystems include unsustainable exploitation of marine resources (for example overfishing of certain species), marine pollution, climate change, and building on coastal areas.

## Freshwater

Freshwater ecosystem.

Freshwater ecosystems cover 0.80% of the Earth's surface and inhabit 0.009% of its total water. They generate nearly 3% of its net primary production. Freshwater ecosystems contain 41% of the world's known fish species.

There are three basic types of freshwater ecosystems:

- Lentic: slow moving water, including pools, ponds, and lakes.

- Lotic: faster moving water, for example streams and rivers.

- Wetlands: areas where the soil is saturated or inundated for at least part of the time.

## Lentic

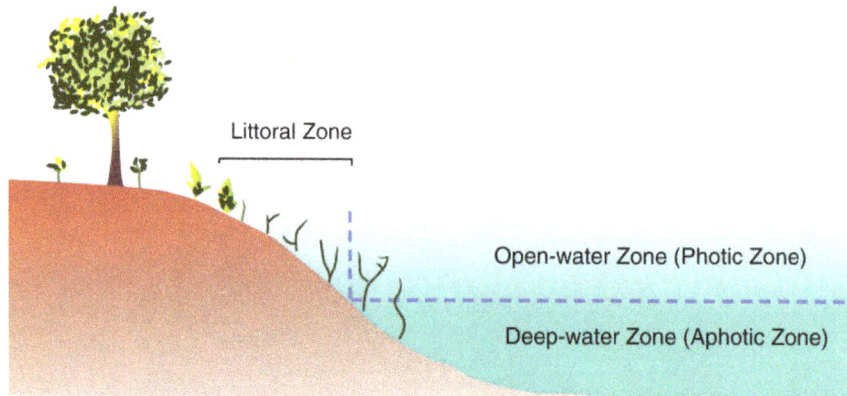

The three primary zones of a lake.

Lake ecosystems can be divided into zones. One common system divides lakes into three zones (see figure). The first, the littoral zone, is the shallow zone near the shore. This is where rooted wetland plants occur. The offshore is divided into two further zones, an open water zone and a deep water zone. In the open water zone (or photic zone) sunlight supports photosynthetic algae, and the species that feed upon them. In the deep water zone, sunlight is not available and the food web is based on detritus entering from the littoral and photic zones. Some systems use other names. The off shore areas may be called the pelagic zone, the photic zone may be called the limnetic zone and the aphotic zone may be called the profundal zone. Inland from the littoral zone one can also frequently identify a riparian zone which has plants still affected by the presence of the lake—this can include effects from windfalls, spring flooding, and winter ice damage. The production of the lake as a whole is the result of production from plants growing in the littoral zone, combined with production from plankton growing in the open water.

Wetlands can be part of the lentic system, as they form naturally along most lake shores, the width of the wetland and littoral zone being dependent upon the slope of the shoreline and the amount of natural change in water levels, within and among years. Often dead trees accumulate in this zone, either from windfalls on the shore or logs transported to the site during floods. This woody debris provides important habitat for fish and nesting birds, as well as protecting shorelines from erosion.

Two important subclasses of lakes are ponds, which typically are small lakes that intergrade with wetlands, and water reservoirs. Over long periods of time, lakes, or bays within them, may gradually become enriched by nutrients and slowly fill in with organic sediments, a process called succession. When humans use the watershed, the volumes of sediment entering the lake can accelerate this process. The addition of sediments and nutrients to a lake is known as eutrophication.

## Ponds

Ponds are small bodies of freshwater with shallow and still water, marsh, and aquatic plants. They can be further divided into four zones: vegetation zone, open water, bottom mud and surface film. The size and depth of ponds often varies greatly with the time of year; many ponds are produced by spring flooding from rivers. Food webs are based both on free-floating algae and upon aquatic plants. There is usually a diverse array of aquatic life, with a few examples including algae, snails, fish, beetles, water bugs, frogs, turtles, otters and muskrats. Top predators may include large fish, herons, or alligators. Since fish are a major predator upon amphibian larvae, ponds that dry up each year, thereby killing resident fish, provide important refugia for amphibian breeding. Ponds that dry up completely each year are often known as vernal pools. Some ponds are produced by animal activity, including alligator holes and beaver ponds, and these add important diversity to landscapes.

## Lotic

The major zones in river ecosystems are determined by the river bed's gradient or by the velocity of the current. Faster moving turbulent water typically contains greater concentrations of dissolved oxygen, which supports greater biodiversity than the slow moving water of pools. These distinctions form the basis for the division of rivers into upland and lowland rivers. The food base of streams within riparian forests is mostly derived from the trees, but wider streams and those that lack a canopy derive the majority of their food base from algae. Anadromous fish are also an important source of nutrients. Environmental threats to rivers include loss of water, dams, chemical pollution and introduced species. A dam produces negative effects that continue down the watershed. The most important negative effects are the reduction of spring flooding, which damages wetlands, and the retention of sediment, which leads to loss of deltaic wetlands.

## Wetlands

Wetlands are dominated by vascular plants that have adapted to saturated soil. There are four main types of wetlands: swamp, marsh, fen and bog (both fens and bogs are types of mire). Wetlands are the most productive natural ecosystems in the world because of the proximity of water and soil. Hence they support large numbers of plant and animal species. Due to their productivity, wetlands are often converted into dry land with dykes and drains and used for agricultural purposes. The construction of dykes, and dams, has negative consequences for individual wetlands and entire watersheds. Their closeness to lakes and rivers means that they are often developed for human settlement. Once settlements are constructed and protected by dykes, the settlements then become vulnerable to land subsidence and ever increasing risk of flooding. The Louisiana coast around New Orleans is a well-known example; the Danube Delta in Europe is another.

## Functions

Aquatic ecosystems perform many important environmental functions. For example, they recycle nutrients, purify water, attenuate floods, recharge ground water and provide habitats for wildlife. Aquatic ecosystems are also used for human recreation, and are very important to the tourism industry, especially in coastal regions.

The health of an aquatic ecosystem is degraded when the ecosystem's ability to absorb a stress has been exceeded. A stress on an aquatic ecosystem can be a result of physical, chemical or biological alterations of the environment. Physical alterations include changes in water temperature, water flow and light availability. Chemical alterations include changes in the loading rates of biostimulatory nutrients, oxygen consuming materials, and toxins. Biological alterations include over-harvesting of commercial species and the introduction of exotic species. Human populations can impose excessive stresses on aquatic ecosystems. There are many examples of excessive stresses with negative consequences. Consider three. The environmental history of the Great Lakes of North America illustrates this problem, particularly how multiple stresses, such as water pollution, over-harvesting and invasive species can combine. The Norfolk Broadlands in England illustrate similar decline with pollution and invasive species. Lake Pontchartrain along the Gulf of Mexico illustrates the negative effects of different stresses including levee construction, logging of swamps, invasive species and salt water intrusion.

## Abiotic Characteristics

An ecosystem is composed of biotic communities that are structured by biological interactions and abiotic environmental factors. Some of the important abiotic environmental factors of aquatic ecosystems include substrate type, water depth, nutrient levels, temperature, salinity, and flow. It is often difficult to determine the relative importance of these factors without rather large experiments. There may be complicated feedback loops. For example, sediment may determine the presence of aquatic plants, but aquatic plants may also trap sediment, and add to the sediment through peat.

The amount of dissolved oxygen in a water body is frequently the key substance in determining the extent and kinds of organic life in the water body. Fish need dissolved oxygen to survive, although their tolerance to low oxygen varies among species; in extreme cases of low oxygen some fish even resort to air gulping. Plants often have to produce aerenchyma, while the shape and size of leaves may also be altered. Conversely, oxygen is fatal to many kinds of anaerobic bacteria.

Nutrient levels are important in controlling the abundance of many species of algae. The relative abundance of nitrogen and phosphorus can in effect determine which species of algae come to dominate. Algae are a very important source of food for aquatic life, but at the same time, if they become over-abundant, they can cause declines in fish when they decay. Similar over-abundance of algae in coastal environments such as the Gulf of Mexico produces, upon decay, a hypoxic region of water known as a dead zone.

The salinity of the water body is also a determining factor in the kinds of species found in the water body. Organisms in marine ecosystems tolerate salinity, while many freshwater organisms are intolerant of salt. The degree of salinity in an estuary or delta is an important control upon the type of wetland (fresh, intermediate, or brackish), and the associated animal species. Dams built upstream may reduce spring flooding, and reduce sediment accretion, and may therefore lead to saltwater intrusion in coastal wetlands.

Freshwater used for irrigation purposes often absorbs levels of salt that are harmful to freshwater organisms.

## Biotic Characteristics

The biotic characteristics are mainly determined by the organisms that occur. For example, wetland plants may produce dense canopies that cover large areas of sediment—or snails or geese may graze the vegetation leaving large mud flats. Aquatic environments have relatively low oxygen levels, forcing adaptation by the organisms found there. For example, many wetland plants must produce aerenchyma to carry oxygen to roots. Other biotic characteristics are more subtle and difficult to measure, such as the relative importance of competition, mutualism or predation. There are a growing number of cases where predation by coastal herbivores including snails, geese and mammals appears to be a dominant biotic factor.

## Autotrophic Organisms

Autotrophic organisms are producers that generate organic compounds from inorganic material. Algae use solar energy to generate biomass from carbon dioxide and are possibly the most important autotrophic organisms in aquatic environments. Of course, the more shallow the water, the greater the biomass contribution from rooted and floating vascular plants. These two sources combine to produce the extraordinary production of estuaries and wetlands, as this autotrophic biomass is converted into fish, birds, amphibians and other aquatic species.

Chemosynthetic bacteria are found in benthic marine ecosystems. These organisms are able to feed on hydrogen sulfide in water that comes from volcanic vents. Great concentrations of animals that feed on these bacteria are found around volcanic vents. For example, there are giant tube worms (*Riftia pachyptila*) 1.5 m in length and clams (*Calyptogena magnifica*) 30 cm long.

## Heterotrophic Organisms

Heterotrophic organisms consume autotrophic organisms and use the organic compounds in their bodies as energy sources and as raw materials to create their own biomass. Euryhaline organisms are salt tolerant and can survive in marine ecosystems, while stenohaline or salt intolerant species can only live in freshwater environments.

# Aquaculture of Salmonids

The aquaculture of salmonids is the farming and harvesting of salmonids under controlled conditions for both commercial and recreational purposes. Salmonids (particularly salmon and steelhead), along with carp, are the two most important fish groups in aquaculture. The most commonly commercially farmed salmonid is the Atlantic salmon. In the U.S. Chinook salmon and rainbow trout are the most commonly farmed salmonids for recreational and subsistence fishing through the National Fish Hatchery System. In Europe, brown trout are the most commonly reared fish for recreational restocking. Commonly farmed non-salmonid fish groups include tilapia, catfish, sea bass and bream.

In 2007 the aquaculture of salmonids was worth US$10.7 billion globally. Salmonid aquaculture production grew over ten-fold during the 25 years from 1982 to 2007. Leading producers of farmed salmonids are Norway with 33 percent, Chile with 31 percent, and other European producers with 19 percent.

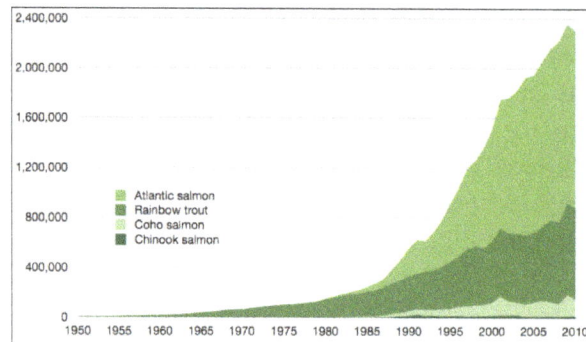

Aquaculture production of salmonids in tonnes
1950–2010 as reported by the FAO

There is currently much controversy about the ecological and health impacts of intensive salmonids aquaculture. There are particular concerns about the impacts on wild salmon and other
marine life. Some of this controversy is part of a major commercial competitive fight for market
share and price between Alaska commercial salmonid fishermen and the rapidly evolving salmonid aquaculture industry.

Salmon farm in the archipelago of Finland

## Methods

The aquaculture or farming of salmonids can be contrasted with capturing wild salmonids using
commercial fishing techniques. However, the concept of "wild" salmon as used by the Alaska Seafood Marketing Institute includes stock enhancement fish produced in hatcheries that have historically been considered ocean ranching. The percentage of the Alaska salmon harvest resulting
from ocean ranching depends upon the species of salmon and location, however it is all marketed
as "wild Alaska salmon".

Methods of salmonid aquaculture originated in late 18th century fertilization trials in Europe. In the
late 19th century, salmon hatcheries were used in Europe and North America. From the late 1950s,
enhancement programs based on hatcheries were established in the United States, Canada, Japan and
the USSR. The contemporary technique using floating sea cages originated in Norway in the late 1960s.

Assynt Salmon hatchery, near Inchnadamph in the Scottish Highlands.

Salmonids are usually farmed in two stages and in some places maybe more. First, the salmon are hatched from eggs and raised on land in freshwater tanks. Increasing the accumulated thermal units of water during incubation reduces time to hatching. When they are 12 to 18 months old, the smolt (juvenile salmon) are transferred to floating sea cages or net pens anchored in sheltered bays or fjords along a coast. This farming in a marine environment is known as mariculture. There they are fed pelleted feed for another 12 to 24 months, when they are harvested.

Very young fertilised salmon eggs; notice the developing eyes and vertebral column.

Norway produces 33 percent of the world's farmed salmonids, and Chile produces 31 percent. The coastlines of these countries have suitable water temperatures and many areas well protected from storms. Chile is close to large forage fisheries which supply fish meal for salmon aquaculture. Scotland and Canada are also significant producers.

Modern salmonid farming systems are intensive. Their ownership is often under the control of huge agribusiness corporations, operating mechanized assembly lines on an industrial scale. In 2003, nearly half of the world's farmed salmon was produced by just five companies.

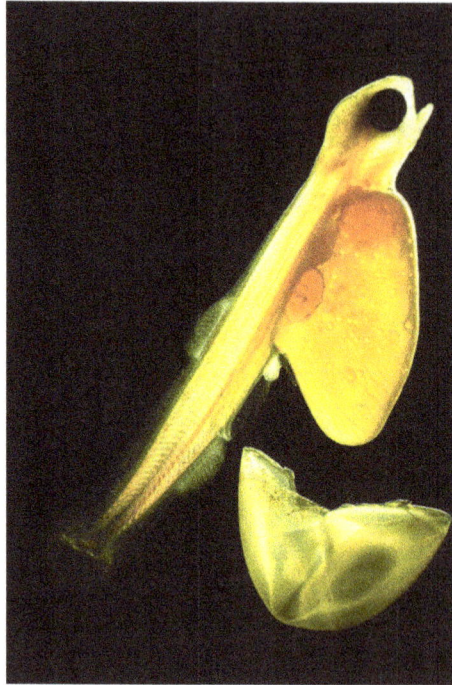

Salmon egg hatching. In about 24hrs it will be a fry without the yolk sac.

## Hatcheries

Modern commercial hatcheries for supplying salmon smolts to aquaculture net pens have been shifting to Recirculating Aquaculture Systems (RAS) where the water is recycled within the hatchery. This allows location of the hatchery to be independent of a significant fresh water supply and allows economical temperature control to both speed up and slow down the growth rate to match the needs of the net pens.

Conventional hatchery systems operate flow through where spring water or other water source flow into the hatchery. The eggs are then hatched in trays and the salmon smolts produced in raceways. The waste products from the growing salmon fry and the feed are usually discharged into the local river. Conventional flow through hatcheries, for example the majority of Alaska's enhancement hatcheries, use more than 100 tonnes (16,000 st) of water to produce a kg of smolts.

An alternative method to hatching in freshwater tanks is to use spawning channels. These are artificial streams, usually parallel to an existing stream with concrete or rip-rap sides and gravel bottoms. Water from the adjacent stream is piped into the top of the channel, sometimes via a header pond to settle out sediment. Spawning success is often much better in channels than in adjacent streams due to the control of floods which in some years can wash out the natural redds (pronounced same as the color 'red'). Because of the lack of floods, spawning channels must sometimes be cleaned out to remove accumulated sediment. The same floods which destroy natural redds also clean them out. Spawning channels preserve the natural selection of natural streams as there is no temptation, as in hatcheries, to use prophylactic chemicals to control diseases. However, exposing fish to wild parasites and pathogens using uncontrolled water supplies, combined with the high cost of spawning channels, makes this technology unsuitable for salmon aquaculture businesses. This type of technology is only useful for stock enhancement programs.

## Sea Cages

Sea cages, also called sea pens or net pens, are usually made of mesh framed with steel or plastic. They can be square or circular, 10 to 32 metres (33 to 105 ft) across and 10 metres (33 ft) deep, with volumes between 1,000 to 10,000 cubic metres (35,000 to 353,000 cu ft). A large sea cage can house up to 90,000 fish.

They are usually placed side by side to form a system called a *seafarm* or *seasite*, with a floating wharf and walkways along the net boundaries. Additional nets can also surround the seafarm to keep out predatory marine mammals. Stocking densities range from 8 to 18 kilograms (18 to 40 lb) per cubic metre for Atlantic salmon and 5 to 10 kilograms (5.0 to 10.0 kg) per cubic metre for Chinook salmon.

In contrast to closed or recirculating systems, the open net cages of salmonid farming lower production costs, but provide no effective barrier to the discharge of wastes, parasites and disease into the surrounding coastal waters. Farmed salmon in open net cages can escape into wild habitats, for example, during storms.

An emerging wave in aquaculture is applying the same farming methods used for salmonids to other carnivorous finfish species, such as cod, bluefin tuna, halibut and snapper. However, this is likely to have the same environmental drawbacks as salmon farming.

A second emerging wave in aquaculture is the development of copper alloys as netting materials. Copper alloys have become important netting materials because they are antimicrobial (i.e., they destroy bacteria, viruses, fungi, algae, and other microbes) and they therefore prevent biofouling (i.e., the undesirable accumulation, adhesion, and growth of microorganisms, plants, algae, tubeworms, barnacles, mollusks, and other organisms). By inhibiting microbial growth, copper alloy aquaculture cages avoid costly net changes that are necessary with other materials. The resistance of organism growth on copper alloy nets also provides a cleaner and healthier environment for farmed fish to grow and thrive.

## Feeding

Salmonids are carnivorous and are currently being fed compound fish feeds containing fish meal and other feed ingredients, ranging from wheat byproducts to soybean meal and feather meal. Being aquatic carnivores, salmonids don't tolerate or properly metabolize many plant based carbohydrates and use fats instead of carbohydrates as a primary energy source.

With the amount of worldwide fish meal production being almost a constant amount for the last 30+ years and at maximum sustainable yield (MSY), much of the fish meal market has shifted from chicken and pig feed to fish and shrimp feeds as aquaculture has grown in this time period.

Work continues on substituting vegetable proteins and protein concentrates for fish meal in the salmonid diet. As of 2014, an enzymatic process can be used to lower the carbohydrate content of barley, making it a high-protein fish feed suitable for salmon. Many other substitutions for fish meal are known, and diets containing zero fish meal are possible. For example, a planned closed containment salmon fish farm in Scotland uses ragworms, algae and amino acids as feed. However, commercial economic animal diets are determined by least cost linear programming models that are effectively competing with similar models for chicken and pig feeds for the same feed ingredients and these models show that fish meal is more useful in aquatic diets than in chicken diets, where they can make the chickens taste like fish. Unfortunately, this substitution can result in lower levels of the highly valued omega-3 content in the farmed product. However, when vegetable oil is used in the growing diet as an energy source and a different finishing diet containing high omega-3 content fatty acids from either fish oil, algae oils or some vegetable oils are used a few months before harvest, this problem is eliminated.

At the present time, more than 50 percent of the world fish oil production is fed to farmed salmonids.

Farm raised salmonids are also fed the carotenoids astaxanthin and canthaxanthin, so that their flesh colour matches wild salmon, which also contain the same carotenoid pigments from their diet in the wild.

On a dry-dry basis, it takes 2–4 kg of wild caught fish to produce one kg of salmon. Wild salmon require about 10 kg of forage fish to produce a kg of salmon, as part of the normal trophic level energy transfer. The difference between the two numbers is related to farmed salmon feed containing other ingredients beyond fish meal and the fact that farmed fish do not expend energy hunting.

## Harvesting

Modern harvesting methods are shifting towards using wet well ships to transport live salmon to the processing plant. This allows the fish to be killed, bled, and filleted before rigor has occurred. This results in superior product quality to the customer along with more humane processing. To obtain maximum quality, it is necessary to minimize the level of stress in the live salmon until actually being electrically and percussively killed and the gills slit for bleeding. These improvements in processing time and freshness to the final customer are commercially significant and forcing the commercial wild fisheries to upgrade their processing to the benefit of all seafood consumers.

An older method of harvesting is to use a sweep net, which operates a bit like a purse seine net. The sweep net is a big net with weights along the bottom edge. It is stretched across the pen with the bottom edge extending to the bottom of the pen. Lines attached to the bottom corners are raised, herding some fish into the purse, where they are netted. Before killing, the fish are usually rendered unconscious in water saturated in carbon dioxide, although this practice is being phased out in some countries due to ethical and product quality concerns. More advanced systems use a percussive-stun harvest system that kills the fish instantly and humanely with a blow to the head from a pneumatic piston. They are then bled by cutting the gill arches and immediately immersing them in iced water. Harvesting and killing methods are designed to minimise scale loss, and avoid the fish releasing stress hormones, which negatively affect flesh quality.

## Wild Versus Farmed

Wild salmonids are captured from wild habitats using commercial fishing techniques. Most wild salmonids are caught in North American, Japanese and Russian fisheries. The following table shows the changes in production of wild salmonids and farmed salmonids over a period of 25 years, as reported by the FAO. Russia, Japan and Alaska all operate major hatchery based stock enhancement programs that are really ocean ranching. The resulting fish hatchery fish are defined as "wild" for FAO and marketing purposes.

| Salmonid production in tonnes by species | | | | | |
|---|---|---|---|---|---|
| | 1982 | | 2007 | | 2013 |
| Species | Wild | Farmed | Wild | Farmed | |
| Atlantic salmon | 10,326 | 13,265 | 2,989 | 1,433,708 | 2,087,110 |
| Steelhead | | 171,946 | | 604,695 | |
| Coho salmon | 42,281 | 2,921 | 17,200 | 115,376 | |
| Chinook salmon | 25,147 | | 8,906 | 11,542 | |
| Pink salmon | 170,373 | | 495,986 | | |
| Chum salmon | 182,561 | | 303,205 | | |
| Sockeye salmon | 128,176 | | 164,222 | | |

| Total salmonid production | | | | |
|---|---|---|---|---|
| | 1982 | | 2007 | |
| | tonnes | percent | tonnes | percent |
| Wild | 558,864 | 75 | 992,508 | 31 |
| Farmed | 188,132 | 25 | 2,165,321 | 69 |
| Overall | 746,996 | | 3,157,831 | |

## Issues

There is currently much controversy about the ecological and health impacts of intensive salmonid aquaculture. There are particular concerns about the impacts on wild salmonids and other marine life and on the incomes of commercial salmonid fishermen.

## Disease and Parasites

In 1972, *Gyrodactylus*, a monogenean parasite, was introduced with live trout and salmon from Sweden (Baltic stocks are resistant to it) into government operated hatcheries in Norway. From the hatcheries, infected eggs, smolt and fry was implanted in many rivers with the goal to strengthen the wild salmon stocks, but caused instead devastation to some of the wild salmon populations affected.

In 1984, infectious salmon anemia (ISAv) was discovered in Norway in an Atlantic salmon hatchery. Eighty percent of the fish in the outbreak died. ISAv, a viral disease, is now a major threat to the viability of Atlantic salmon farming. It is now the first of the diseases classified on List One of the European Commission's fish health regime. Amongst other measures, this requires the total eradication of the entire fish stock should an outbreak of the disease be confirmed on any farm. ISAv seriously affects salmon farms in Chile, Norway, Scotland and Canada, causing major economic losses to infected farms. As the name implies, it causes severe anemia of infected fish. Unlike mammals, the red blood cells of fish have DNA, and can become infected with viruses. The fish develop pale gills, and may swim close to the water surface, gulping for air. However, the disease can also develop without the fish showing any external signs of illness, the fish maintain a normal appetite, and then they suddenly die. The disease can progress slowly throughout an infected farm and, in the worst cases, death rates may approach 100 percent. It is also a threat to the dwindling stocks of wild salmon. Management strategies include developing a vaccine and improving genetic resistance to the disease.

In the wild, diseases and parasites are normally at low levels, and kept in check by natural predation on weakened individuals. In crowded net pens they can become epidemics. Diseases and parasites also transfer from farmed to wild salmon populations. A recent study in British Columbia links the spread of parasitic sea lice from river salmon farms to wild pink salmon in the same river. The European Commission (2002) concluded "The reduction of wild salmonid abundance is also linked to other factors but there is more and more scientific evidence establishing a direct link between the number of lice-infested wild fish and the presence of cages in the same estuary." It is reported that wild salmon on the west coast of Canada are being driven to extinction by sea lice from nearby salmon farms. These predictions have been disputed by other scientists and recent harvests have indicated that the predictions were in error. In 2011, Scotttish salmon farming introduced the use of farmed wrasse for the purpose of cleaning farmed salmon of ectoparasites.

## Pollution and Contaminants

Salmonid farms are typically sited in marine ecosystems with good water quality, high water exchange rates, current speeds fast enough to prevent pollution of the bottom but slow enough to prevent pen damage, protection from major storms, reasonable water depth and a reasonable distance from major infrastructure such as ports, processing plants and logistical facilities like airports. Logistical considerations are significant and feed and maintenance labor must be transported to the facility and the product returned. Siting decisions are complicated by complex politically driven permitting problems in many countries that prevents optimal locations for the farms.

In sites without adequate currents there can be an accumulation of heavy metals on the benthos (seafloor) near the salmon farms, particularly copper and zinc.

Contaminants are commonly found in the flesh of farmed and wild salmon yet seldom exceed tolerance levels set by health authorities. A 2004 study, reported in *Science*, analysed farmed and wild salmon for organochlorine contaminants. They found the contaminants were higher in farmed salmon. Within the farmed salmon, European (particularly Scottish) salmon had the highest levels, and Chilean salmon the lowest. The FDA and Health Canada have established a tolerance/limit for PCBs in commercial fish of 2000 ppb A follow up study confirmed this, and found levels of dioxins, chlorinated pesticides, PCBs and other contaminants up to ten times greater in farmed salmon than wild Pacific salmon. On a positive note, further research using the same fish samples used in the previous study, showed that farmed salmon contained levels of beneficial fatty acids that were two to three times higher than wild salmon. A follow up benefit-risk analysis on salmon consumption balanced the cancer risks with the (n–3) fatty acid advantages of salmon consumption. It is for this reason that current methods for this type of analysis take into consideration the lipid content of the sample in question PCBs specifically are lipophyllic therefore found in higher concentrations in fattier fish in general thus the higher level of PCB in the farmed fish is in relation to the higher content of beneficial n–3 and n–6 lipids they contain. They found that recommended levels of (n–3) fatty acid consumption can be achieved eating farmed salmon with acceptable carcinogenic risks, but recommended levels of EPA+DHA intake cannot be achieved solely from farmed (or wild) salmon without unacceptable carcinogenic risks. The conclusions of this paper from 2005 were that

"...consumers should not eat farmed fish from Scotland, Norway and eastern Canada more than three times a year; farmed fish from Maine, western Canada and Washington state no more than three to six times a year; and farmed fish from Chile no more than about six times a year. Wild chum salmon can be consumed safely as often as once a week, pink salmon, Sockeye and Coho about twice a month and Chinook just under once a month."

A research paper from 2008 titled "Balancing the risks and benefits of fish for sensitive populations" contradicts the above recommendation in light of the fact that the levels of all in that study were on average 100 times below that set as maximum by the FDA, CIA, and EFSA and any risk posed by these contaminants is far outweighed by the proven benefits of eating farmed or wild salmon Due to this fact: Health Canada currently believes that there is no need for specific advice regarding fish consumption vis-à-vis PCB exposure.

Current Canadian dietary guidelines state Eat at least two Food Guide Servings of fish each week. Choose fish such as char, herring, mackerel, salmon, sardines and trout.

The US in their Dietary guidelines for 2010 recommends eating 8 ounces per week of a variety of seafood and 12 ounces for lactating mothers. No upper limits set and no restrictions on eating farmed or wild salmon.

In an "Update of the monitoring of levels of dioxins and PCBs in food and feed" to the European Food Safety Authority in July 2012 stated unequivocally

"Farmed salmon and trout contained on average less dioxins and PCBs than wild-caught salmon and trout."

This quote is from the European Food information Council (EUFIC) in reaction to the 2004 paper "Global Assessment of Organic Contaminants in Farmed Salmon"

"Public concerns were raised earlier this year following the publication of a study by US researchers, who suggested that the levels of organic pollutants, including dioxins and PCBs, in farmed salmon could pose a health risk. Their advice to consume less than one half portion of farmed salmon (from specific areas) per month was in direct contrast to advice from food authorities to eat one portion of oily fish per week. This study did not, however, present new data as levels of contaminants were consistent with those previously reported in smaller studies and remained within internationally accepted safety guidelines. The discrepancy arose because the authors based their advice on a method of risk analysis that is not internationally accepted by toxicologists and other food safety experts. Food safety authorities in Europe and in the USA agreed that the study did not raise new health concerns and that eating one portion of farmed salmon per week was still considered safe." and was followed by these words of advice "The consumer's decision to include or exclude any food from the diet should be based on informed science rather than media headlines."

## Impact on Wild Salmonids

Farmed salmonids can, and often do, escape from sea cages. If the farmed salmonid is not native, it can compete with native wild species for food and habitat. If the farmed salmonid is native, it can interbreed with the wild native salmonids. Such interbreeding can reduce genetic diversity, disease resistance and adaptability. In 2004, about 500,000 salmon and trout escaped from ocean net pens off Norway. Around Scotland, 600,000 salmon were released during storms. Commercial fishermen targeting wild salmon not infrequently catch escaped farm salmon. At one stage, in the Faroe Islands, 20 to 40 percent of all fish caught were escaped farm salmon.

Sea lice, particularly *Lepeophtheirus salmonis* and various *Caligus* species, including *Caligus clemensi* and *Caligus rogercresseyi*, can cause deadly infestations of both farm-grown and wild salmon. Sea lice are naturally occurring and abundant ectoparasites which feed on mucus, blood, and skin, and migrate and latch onto the skin of salmon during planktonic *nauplii* and *copepodid* larval stages, which can persist for several days. Large numbers of highly populated, open-net salmon farms can create exceptionally large concentrations of sea lice; when exposed in river estuaries containing large numbers of open-net farms, many young wild salmon are infected, and do not survive as a result. Adult salmon may survive otherwise critical numbers of sea lice, but small, thin-skinned juvenile salmon migrating to sea are highly vulnerable. In 2007, mathematical studies of data available from the Pacific coast of Canada indicated the louse-induced mortality of pink salmon in some regions was over 80 percent. Later that year, in reaction to the 2007 mathematical study mentioned above, Canadian federal fisheries scientists Kenneth Brooks and Simon Jones published a critique titled "Perspectives on Pink Salmon and Sea Lice: Scientific Evidence Fails to Support the Extinction Hypothesis " The time since these studies has shown a general increase in abundance of Pink Salmon in the Broughton Archipelago. Another comment in the scientific literature by Canadian Government Fisheries scientists Brian Riddell and Richard Beamish et al. came to the conclusion that there is no correlation between farmed salmon louse numbers and returns of pink salmon to the Broughton Archipelago. And in relation to the 2007 Krkosek extinction theory:"the data was used selectively and conclusions do not match with recent observations of returning salmon".

A 2008 meta-analysis of available data shows that salmonid farming reduces the survival of associated wild salmonid populations. This relationship has been shown to hold for Atlantic, steelhead, pink, chum, and coho salmon. The decrease in survival or abundance often exceeds 50 percent. However, these studies are all correlation analysis and correlation doesn't equal causation, especially when similar salmon declines were occurring in Oregon and California, which have no salmon aquaculture or marine net pens. Independent of the predictions of the failure of salmon runs in Canada indicated by these studies, the wild salmon run in 2010 was a record harvest.

A 2010 study that made the first use of sea lice count and fish production data from all salmon farms on the Broughton Archipelago found no correlation between the farm lice counts and wild salmon survival. The authors conclude that the 2001 stock collapse was not caused by the farm sea lice population.The study found that the farm sea lice population during the out-migration of juvenile pink salmon was greater in 2000 than that of 2001, but a record salmon escapement in 2001 exonerates sea lice of the year 2002 collapse due to the absence of negative correlation. The authors also note that initial studies had not investigated bacterial and viral causes for the event despite reports of bleeding at the base of the fins, a symptom often associated with infections but not with sea lice exposure under laboratory conditions.

Wild salmon are anadromous. They spawn inland in fresh water and when young migrate to the ocean where they grow up. Most salmon return to the river where they were born, although some stray to other rivers. There is concern about of the role of genetic diversity within salmon runs.

The resilience of the population depends on some fish being able to survive environmental shocks, such as unusual temperature extremes. It is also unclear what the effect of hatchery production has been on the genetic diversity of salmon.

## Genetic Modification

Salmon have been genetically modified in laboratories so they can grow faster. There is opposition to the commercial use of these fish, and, so far, no approval has been given. A Canadian company, Aqua Bounty Farms, has developed a modified Atlantic salmon which grows nearly twice as fast (yielding a fully grown fish at 16–18 months rather than 30), and is more disease resistant and cold tolerant. It also requires 10 percent less food. This was achieved using a chinook salmon gene sequence affecting growth hormones, and a promoter sequence from the ocean pout affecting antifreeze production. Normally, salmon produce growth hormones only in the presence of light. The modified salmon doesn't switch growth hormone production off. The company first submitted the salmon for FDA approval in 1996. A concern with transgenic salmon is what might happen if they escape into the wild. One study, in a laboratory setting, found that modified salmon mixed with their wild cohorts were aggressive in competing, but ultimately failed.

## Impact on Wild Predatory Species

Sea cages can attract a variety of wild predators which can sometimes become entangled in associated netting leading to injury or death. In Tasmania, Australia salmon farming sea cages have entangled white-bellied sea eagles. This has prompted one company, Huon Aquaculture to sponsor a bird rehabilitation centre and trial more robust netting.

## Impact on Forage Fish

The use of forage fish for fish meal production has been almost a constant for the last thirty years and at the maximum sustainable yield, while the market for fish meal has shifted from chicken, pig and pet food to aquaculture diets. The fact that this market shift at constant production is an economic decision having no impact on the forage fish harvest rates for fish meal implies that the development of salmon aquaculture had no impact on forage fish harvest rates.

Fish do not actually produce omega-3 fatty acids, but instead accumulate them from either consuming microalgae that produce these fatty acids, as is the case with forage fish like herring and sardines, or, as is the case with fatty predatory fish, like salmon, by eating prey fish that have accumulated omega-3 fatty acids from microalgae. To satisfy this requirement, more than 50 percent of the world fish oil production is fed to farmed salmon.

In addition, salmon require nutritional intakes of protein, protein which is often supplied to them in the form of fish meal as the lowest cost alternative protein. Consequently, farmed salmon consume more fish than they generate as a final product.

## Salmon Aquaculture Dialogue

In 2004 the World Wide Fund for Nature (WWF) initiated the *Salmon Aquaculture Dialogue.*

The aim of the dialogue is to produce an environmental standard for farmed salmon by 2010. The WWF have identified what they call "seven key environmental and social impacts", which they characterise as follows

1.  Benthic impacts and siting: Chemicals and excess nutrients from food and feces associated with salmon farms can disturb the flora and fauna on the ocean bottom (benthos).

2.  Chemical inputs: Excessive use of chemicals - such as antibiotics, anti-foulants and pesticides - or the use of banned chemicals can have unintended consequences for marine organisms and human health.

3.  Disease/parasites: Viruses and parasites can transfer between farmed and wild fish, as well as among farms.

4.  Escapes: Escaped farmed salmon can compete with wild fish and interbreed with local wild stocks of the same population, altering the overall pool of genetic diversity.

5.  Feed: A growing salmon farming business must control and reduce its dependency upon fishmeal and fishoil - a primary ingredient in salmon feed—so as not to put additional pressure on the world's fisheries. Fish caught to make fishmeal and oil currently represent one-third of the global fish harvest.

6.  Nutrient loading and carrying capacity: Excess food and fish waste in the water have the potential to increase the levels of nutrients in the water. This can cause the growth of algae, which consumes oxygen that is meant for other plant and animal life.

7.  Social issues: Salmon farming often employs a large number of workers on farms and in processing plants, potentially placing labor practices and worker rights under public scrutiny. Additionally, conflicts can arise among users of the shared coastal environment."

— *World Wide Fund for Nature,*

## Hatch and Release

Another form of salmon production, which is safer but less controllable, is to raise salmon in hatcheries until they are old enough to become independent. They are then released into rivers, often in an attempt to increase the salmon population. This practice was very common in countries like Sweden before the Norwegians developed salmon farming, but is seldom done by private companies, as anyone may catch the salmon when they return to spawn, limiting a company's chances of benefiting financially from their investment. Because of this, the method has mainly been used by various public authorities and non profit groups like the Cook Inlet Aquaculture Association as a way of artificially increasing salmon populations in situations where they have declined due to overharvest, construction of dams, and habitat destruction or disruption. Unfortunately, there can be negative consequences to this sort of population manipulation, including genetic "dilution" of the wild stocks, and many jurisdictions are now beginning to discourage supplemental fish planting in favour of harvest controls and habitat improvement and protection. A variant method of fish stocking, called ocean ranching, is under development in Alaska. There, the young salmon are released into the ocean far from any wild salmon streams. When it is time for them to spawn, they return to where they were released where fishermen can then catch them.

## Species

### Atlantic Salmon

In their natal streams, Atlantic salmon are considered a prized recreational fish, pursued by avid fly anglers during its annual runs. At one time, the species supported an important commercial fishery and a supplemental food fishery. However, the wild Atlantic salmon fishery is commercially dead; after extensive habitat damage and overfishing, wild fish make up only 0.5 percent of the Atlantic salmon available in world fish markets. The rest are farmed, predominantly from aquaculture in Chile, Canada, Norway, Russia, the UK and Tasmania in Australia.

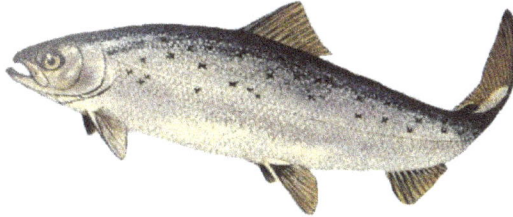

Atlantic salmon

Atlantic salmon is, by far, the species most often chosen for farming. It is easy to handle, it grows well in sea cages, commands a high market value and it adapts well to being farmed away from its native habitats.

Adult male and female fish are anesthetized. Eggs and sperm are "stripped", after the fish are cleaned and cloth dried. Sperm and eggs are mixed, washed, and placed into fresh water. Adults recover in flowing, clean, well aerated water. Some researchers have studied cryopreservation of the eggs.

Fry are generally reared in large freshwater tanks for 12 to 20 months. Once the fish have reached the smolt phase, they are taken out to sea where they are held for up to two years. During this time the fish grow and mature in large cages off the coasts of Canada, the United States, or parts of Europe. Generally, cages are made of two nets; inner nets, which wrap around the cages, hold the salmon while outer nets, which are held by floats, keep predators out.

Many Atlantic salmon escape from cages at sea. Those salmon who further breed tend to lessen the genetic diversity of the species leading to lower survival rates, and lower catch rates. On the West Coast of Northern America, the non-native salmon could be an invasive threat, especially in Alaska and parts of Canada. This could cause them to compete with native salmon for resources. Extensive efforts are underway to prevent escapes and the potential spread of Atlantic salmon in the Pacific and elsewhere. The risk of Atlantic Salmon becoming a legitimate invasive threat on the Pacific Coast of N. America is questionable in light of both Canadian and American governments deliberately introducing this species by the millions for a 100-year period starting in the 1900s. Despite these deliberate attempts to establish this species on the pacific coast; there have been no established populations to report.

In 2007, 1,433,708 tonnes of Atlantic salmon were harvested worldwide with a value of $7.58 billion.

## Steelhead

Rainbow trout

In 1989 steelhead were re-classified into the Pacific trout as *Oncorhynchus mykiss* from the former bi-nominals of *Salmo gairdneri* (Columbia River redband trout) and *S. irideus* (coastal rainbow trout). Steelhead are an anadromous form of rainbow trout that migrates between lakes and rivers and the ocean, and are also known as steelhead salmon or ocean trout.

Male ocean phase steelhead salmon

Steelhead are raised in many countries throughout the world. Since the 1950s production has grown exponentially, particularly in Europe and recently in Chile. Worldwide, in 2007, 604,695 tonnes of farmed Steelhead were harvested with a value of $2.59 billion. The largest producer is Chile. In Chile and Norway, the ocean cage production of steelhead has expanded to supply export markets. Inland production of rainbow trout to supply domestic markets has increased strongly in countries such as Italy, France, Germany, Denmark, and Spain. Other significant producing countries include the United States, Iran, Germany, and the UK. Rainbow trout, including juvenile steelhead in fresh water, routinely feed on larval, pupal and adult forms of aquatic insects (typically caddisflies, stoneflies, mayflies and aquatic diptera). They also eat fish eggs and adult forms of terrestrial insects (typically ants, beetles, grasshoppers and crickets) that fall into the water. Other prey include small fish up to one-third of their length, crayfish, shrimp, and other crustaceans. As rainbow trout grow, the proportion of fish consumed increases in most populations. Some lake-dwelling forms may become planktonic feeders. In rivers and streams populated with other salmonid species, rainbow trout eat varied fish eggs, including those of salmon, brown and cutthroat trout, mountain whitefish and the eggs of other rainbow trout. Rainbows also consume decomposing flesh from carcasses of other fish. Adult steelhead in the ocean feed primarily on other fish, squid and amphipods. Cultured steelhead are fed a diet formulated to closely resemble their natural diet that includes fish meal, fish oil, vitamins and minerals, and the carotenoid Asthaxanthin for pigmentation.

The steelhead is especially susceptible to enteric redmouth disease. There has been considerable research conducted on redmouth disease, as its implications for steelhead farmers are significant. The disease does not affect humans.

## Coho Salmon

Male ocean phase Coho salmon

The Coho salmon is the state animal of Chiba, Japan.

Coho salmon mature after only one year in the sea, so two separate broodstocks (spawners) are needed, alternating each year. Broodfish are selected from the salmon in the seasites and "transferred to freshwater tanks for maturation and spawning".

Worldwide, in 2007, 115,376 tonnes of farmed Coho salmon were harvested with a value of $456 million. Chile, with about 90 percent of world production, is the primary producer with Japan and Canada producing the rest.

## Chinook Salmon

male ocean phase Chinook

In Alaska, Chinook salmon are the state fish, and are known as "king salmon" because of their large size and flavourful flesh. Those from the Copper River in Alaska are particularly known for their colour, rich flavour, firm texture, and high omega-3 oil content. Alaska has a long-standing ban on finfish aquaculture that was enacted in 1989. Alaska Stat. § 16.40.210

Worldwide, in 2007, 11,542 tonnes (1,817,600 st) of farmed Chinook salmon were harvested with a value of $83 million. New Zealand is the largest producer of farmed king salmon, accounting for over half of world production (7,400 tonnes in 2005). Most of the salmon are farmed in the sea (mariculture) using a method sometimes called sea-cage ranching. Sea-cage ranching takes place in large floating net cages, about 25 metres across and 15 metres deep, moored to the sea floor in clean, fast-flowing coastal waters. Smolt (young fish) from freshwater hatcheries are transferred to cages containing several thousand salmon, and remain there for the rest of their life. They are fed fishmeal pellets high in protein and oil.

male freshwater phase Chinook

Chinook salmon are also farmed in net cages placed in freshwater rivers or raceways, using techniques similar to those used for sea-farmed salmon. A unique form of freshwater salmon farming occurs in some hydroelectric canals in New Zealand. A site in Tekapo, fed by fast cold waters from the Southern Alps, is the highest salmon farm in the world, 677 metres (2,221 ft) above sea level.

Before they are killed, cage salmon are sometimes anaesthetised with a herbal extract. They are then spiked in the brain. The heart beats for a time as the animal is bled from its sliced gills. This method of relaxing the salmon when it is killed produces firm, long-keeping flesh. Lack of disease in wild populations and low stocking densities used in the cages means that New Zealand salmon farmers do not use antibiotics and chemicals that are often needed elsewhere.

## Timeline

- 1527: The life history of the Atlantic salmon is described by Hector Boece of the University of Aberdeen, Scotland.

- 1763: Fertilization trials for Atlantic salmon take place in Germany. Later biologists refined these in Scotland and France.

- 1854: Salmon spawing beds and rearing ponds built along the bank of a river by the Dohulla Fishery, Ballyconneely, Ireland.

- 1864: Hatchery raised Atlantic salmon fry were released in the River Plenty, Tasmania in a failed attempt to establish a population in Australia

- 1892: Hatchery raised Atlantic salmon fry were released in the Umkomass river in South Africa in a failed attempt to establish a population in Africa.

- Late 19th century: Salmon hatcheries are used in Europe, North America, and Japan to enhance wild populations.

- 1961: Hatchery raised Atlantic salmon fry were released in the rivers of the Falkland Islands in a failed attempt to establish a population in the South Atlantic.

- Late 1960s: First salmon farms established in Norway and Scotland.

- 1970: Hatchery raised Atlantic salmon fry were released in the rivers of the Kerguelen Islands in a failed attempt to establish a population in the Indian Ocean.

- Early 1970s: Salmon farms established in North America.

- 1975: Gyrodactylus, a small monogenean parasite, spreads from Norwegian hatcheries to wild salmon, probably by means of fishing gear, and devastates some wild salmon populations.

- Late 1970s: Salmon farms established in Chile and New Zealand.

- 1984: Infectious salmon anemia, a viral disease, is discovered in a Norwegian salmon hatchery. Eighty percent of the involved fish die.

- 1985: Salmon farms established in Australia.

- 1987: First reports of escaped Atlantic salmon being caught in wild Pacific salmon fisheries.

- 1988: A storm hits the Faroe Islands releasing millions of Atlantic salmon.

- 1989: Furunculosis, a bacterial disease, spreads through Norwegian salmon farms and wild salmon.

- 1996: World farmed salmon production exceeds wild salmon harvest.

- 2007: A 10-square-mile (26 km²) swarm of Pelagia noctiluca jellyfish wipes out a 100,000 fish salmon farm in Northern Ireland.

# References

- Barange M, Field JG, Harris RP, Eileen E, Hofmann EE, Perry RI and Werner F (2010) Marine Ecosystems and Global Change Oxford University Press. ISBN 978-0-19-955802-5

- Boyd IL, Wanless S and Camphuysen CJ (2006) Top predators in marine ecosystems: their role in monitoring and management Volume 12 of Conservation biology series. Cambridge University Press. ISBN 978-0-521-84773-5

- Christensen V and Pauly D (eds.) (1993) Trophic models of aquatic ecosystems The WorldFish Center, issue 26 of ICLARM Technical Reports, volume 26 of ICLARM conference proceedings. ISBN 9789711022846.

- Davenport J (2008) Challenges to Marine Ecosystems: Proceedings of the 41st European Marine Biology Symposium Volume 202 of Developments in hydrobiology. ISBN 978-1-4020-8807-0

- Levner E, Linkov I and Proth J (2005) Strategic management of marine ecosystems Springer. Volume 50 of NATO Science Series IV. ISBN 978-1-4020-3158-8

- Mann KH and Lazier JRN (2006) Dynamics of marine ecosystems: biological-physical interactions in the oceans Wiley-Blackwell. ISBN 978-1-4051-1118-8

- National Research Council (US) (1996) Freshwater ecosystems: revitalizing educational programs in limnology National Academy Press. ISBN 0-309-05443-5

- Beveridge, Malcolm (1984) Cage and Pen fish farming: Carrying capacity models and environmental impact FAO Fisheries technical paper 255, Rome. ISBN 92-5-102163-5

- Coimbra, João (1 January 2001). Modern Aquaculture in the Coastal Zone: Lessons and Opportunities. IOS Press. pp. 32–. ISBN 978-0-9673355-6-8.

- Harris, Graeme; Milner, Nigel (12 March 2007). Sea Trout: Biology, Conservation and Management. Wiley. pp. 18–. ISBN 978-1-4051-2991-6.

- Knapp G., Roheim C. A. and Anderson J. A. (2007) The Great Salmon Run: Competition between Wild and Farmed Salmon Report of the Institute of Social and Economic Research, University of Alaska Anchorage. ISBN 0-89164-175-0.

- Quinn, Thomas P. (2005). The Behavior and Ecology of Pacific Salmon and Trout. American Fisheries Society. pp. 18–. ISBN 978-0-295-98457-5.

# 5

# Aquaculture Systems

With the growth and establishment of aquaculture as a viable source for marine food, there have evolved different systems of aquaculture practices. Some of these are aquaponics, organic aquaculture, and recirculating aquaculture systems. The major components of aquaculture systems are discussed in this chapter.

## Recirculating Aquaculture System

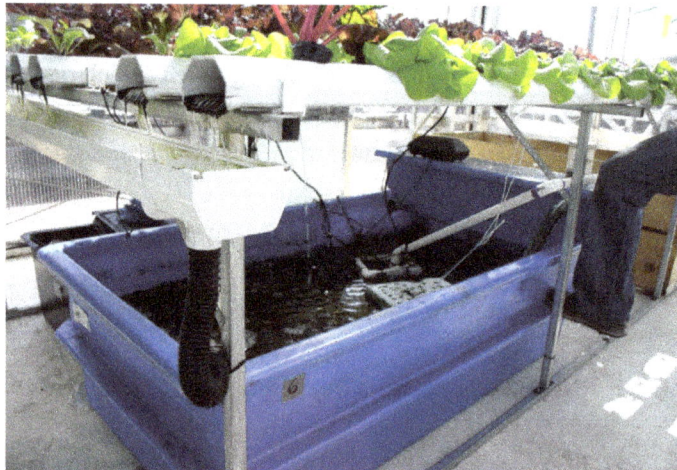

Aquaponics is a type of recirculating aquaculture system where both fish and plants are present.

Recirculating aquaculture systems (RAS) are used in home aquaria and for fish production where water exchange is limited and the use of biofiltration is required to reduce ammonia toxicity. Other types of filtration and environmental control are often also necessary to maintain clean water and provide a suitable habitat for fish. The main benefit of RAS is the ability to reduce the need for fresh, clean water while still maintaining a healthy environment for fish. To be operated economically commercial RAS must have high fish stocking densities, and many researchers are currently conducting studies to determine if RAS is a viable form of intensive aquaculture.

### RAS Water Treatment Processes

A series of treatment processes are utilized to maintain water quality in intensive fish farming operations. These steps are often done in order or sometimes in tandem. After leaving the vessel holding fish the water is first treated for solids before entering a biofilter to convert ammonia, next degassing and oxygenation occur, often followed by heating/cooling and sterilization. Each of theses processes can be completed by using a variety of different methods and equipment, but regardless all must take place to insure a healthy environment that maximizes fish growth and health.

A biofilter and $CO_2$ degasser on an outdoor recirculating aquaculture system used to grow largemouth bass.

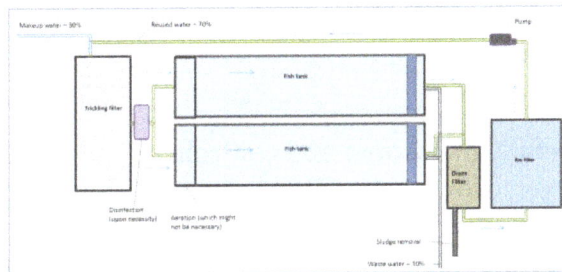

A Recirculating aquaculture system model with 70% degree of recirculation.

## Biofiltration

All RAS relies on biofiltration to convert ammonia ($NH_4^+$ and $NH_3$) excreted by the fish into nitrate. Ammonia is a waste product of fish metabolism and high concentrations (>.02 mg/L) are toxic to most finfish. Nitrifying bacteria are chemoautotrophs that convert ammonia into nitrite then nitrate. A biofilter provides a substrate for the bacterial community, which results in thick biofilm growing within the filter. Water is pumped through the filter, and ammonia is utilized by the bacteria for energy. Nitrate is less toxic than ammonia (>100 mg/L), and can be removed by a denitrifying biofilter or by water replacement. Stable environmental conditions and regular maintenance are required to insure the biofilter is operating efficiently.

## Solids Removal

In addition to treating the liquid waste excreted by fish the solid waste must also be treated, this is done be concentrating and flushing the solids out of the system. Removing solids reduces bacteria growth, oxygen demand, and the proliferation of disease. The simplest method for removing solids is the creation of settling basin where the relative velocity of the water is slow and particles can settle to the bottom of the tank where they are either flushed out or vacuumed out manually using a siphon. However, this method is not viable for RAS operations where a small footprint is desired. Typical RAS solids removal involves a sand filter or particle filter where solids become lodged and can by periodically backflushed out of the filter. Another common method is the use of a mechanical drum filters where water is run over a rotating drum screen that is periodically cleaned by pressurized spray nozzles, and the resulting slurry is treated or sent down the drain. In order to remove extremely fine particles or colloidal solids a protein fractionator may be used with or without the addition of ozone ($O_3$).

## Oxygenation

Reoxygenating the system water is crucial to obtain high production densities. Fish require oxygen to metabolize food and grow, as do bacteria communities in the biofilter. Dissolved oxygen levels can be increased through two methods aeration and oxygenation. In aeration air is pumped through an air stone or similar device that creates small bubbles in the water column, this results in a high surface area where oxygen can dissolve into the water. In general due to slow gas dissolution rates and the high air pressure needed to create small bubbles this method is considered inefficient and the water is instead oxygenated by pumping in pure oxygen. Various methods are used to ensure that during oxygenation all of the oxygen dissolves into the water column. Careful calculation and consideration must be given to the oxygen demand of a given system, and that demand must be met with either oxygenation or aeration equipment.

## pH Control

In all RAS pH must be carefully monitored and controlled. The first step of nitrification in the biofilter consumes alkalinity and lowers the pH of the system. Keeping the pH in a suitable range (5.0-9.0 for freshwater systems) is crucial to maintain the health of both the fish and biofilter. pH is typically controlled by the addition of alkalinity in the form of lime ($CaCO_3$). A low pH will lead to high levels of dissolved carbon dioxide ($CO_2$), which can prove toxic to fish. pH can also be controlled by degassing $CO_2$ in a packed column or with an aerator, this is necessary in intensive systems especially where oxygenation instead of aeration is used in tanks to maintain $O_2$ levels.

## Temperature Control

All fish species have a preferred temperature above and below which that fish will experience negative health effects and eventually death. Warm water species such as Tilapia and Barramundi prefer 24 °C water or warmer, where as cold water species such as trout and salmon prefer water temperature below 16 °C. Temperature also plays an important role in dissolved oxygen (DO) concentrations, with higher temperatures resulting in lower levels of DO. Temperature is controlled through the use of submerged heaters, heat pumps, chillers, and heat exchangers. All four may be used to keep a system operating at the optimal temperature for maximizing fish production.

## Biosecurity

Disease outbreaks occur more readily when dealing with the high fish stocking densities typically employed in intensive RAS. Outbreaks can be reduced by operating multiple independent systems with the same building and isolating water to water contact between systems by cleaning equipment and personnel that move between systems. Also the use of a Ultra Violet (UV) or ozone water treatment system reduces the number of free floating virus and bacteria in the system water. These treatment systems reduce the disease loading that occurs on stressed fish and thus reduces the chance of an outbreak.

# Advantages

Sturgeon grown at high density in a partial recirculating aquaculture system.

- Reduced water requirements as compared to raceway or pond aquaculture systems.

- Reduced land needs due to the high stocking density

- Site selection flexibility and independence from a large, clean water source.

- Reduction in wastewater effluent volume.

- Increased biosecurity and ease in treating disease outbreaks.

- Ability to closely monitor and control environmental conditions to maximize production efficiency. Similarly, independence from weather and variable environmental conditions.

# Disadvantages

- High upfront investment in materials and infrastructure.

- High operating costs mostly due to electricity, and system maintenance.

- A need for highly trained staff to monitor and operate the system.

# Special Types of RAS

## Aquaponics

Combining plants and fish in a RAS is referred to as aquaponics. In this type of system ammonia produced by the fish is not only converted to nitrate but is also removed by the plants from the water. In an aquaponics system fish effectively fertilize the plants, this creates a closed looped system where very little waste is generated and inputs are minimized. Aquaponics provides the advantage of being able to harvest and sell multiple crops.

## Aquariums

Home aquaria and inland commercial aquariums are a form of RAS where the water quality is very carefully controlled and the stocking density of fish is relatively low. In these systems the goal is to display the fish rather than producing food. However, biofilters and other forms of water treatment are still used to reduce the need to exchange water and to maintain water clarity. Just like in traditional RAS water must be removed periodically to prevent nitrate and other toxic chemicals from building up in the system. Coastal aquariums often have high rates of water exchange and are typically not operated as a RAS due to their proximity to a large body of clean water.

# Organic Aquaculture

Organic aquaculture is a holistic method for farming marine species in line with organic principles. The ideals of this practice establish sustainable marine environments with consideration for naturally-occurring ecosystems, use of pesticides, and the treatment of aquatic life. Managing aquaculture organically has become more popular since consumers are concerned about the harmful impacts of aquaculture on themselves and the environment.

Aquaculture is the fastest growing sector of the food system and the availability of certified organic aquaculture products have become more widely available since the mid-1990s. This seafood growing method has become popular in Germany, the United Kingdom and Switzerland, but consumers can be confused or skeptical about the label due to conflicting and misleading standards around the world.

A certified organic product seal on aquaculture products will mean an accredited certifying body has verified that the production methods meet or exceed a country's standard for organic aquaculture production. Organic regulations designed around soil-based systems don't transfer well into aquaculture  and tend to conflict with large-scale, intensive (economically viable) practices/goals. There are a number of problems facing organic aquaculture: difficulty of sourcing and certifying organic juveniles (hatchery or sustainable wild stock); 35-40% higher feed cost; more labour-intensive; time and cost of the certification process; a higher risk of diseases, and uncertain benefits. But, there is a definite consumer demand for organic seafood, and organic aquaculture may become a significant management option with continued research.

## Certification

A number of countries have created their own national standards and certifying bodies for organic aquaculture. While there is not simply one international organic aquaculture standardization process, one of the largest certification organizations is the Global Trust, which delivers assessments and certifications to match the highest quality organic aquaculture standards. The information regarding these standards is available through a personal inquiry.

Many organic aquaculture certifications address a variety of issues including antibiotic and chemical treatments of fish, unrestrained disposal of fish feces into the ocean, fish feeding materials, the habitat of where and how the fish are raised, and proper handling practices including slaughter. Most Organic Aquaculture certifications follow rather strict requirements and standards. These rules may vary between different countries or certification bodies. This leads to confusion when products are imported from other countries, which can result in a backlash from consumers (for example, the Pure Salmon Campaign ).

Defining acceptable practices is also complicated by the variety of species - freshwater, salt-water, shellfish, finfish, mollusks and aquatic plants. The difficulty of screening pollutants out of an aquatic medium, controlling the food supplies and of keeping track of individual fish may mean that fish and shellfish stocks should not be classified as 'livestock' at all under regulations. This point further exemplifies the need for widespread aquaculture certification standard.

## Challenges and Controversy

There is some controversy over licensing restrictions, as some seafood companies propose that wild caught fish should be classified as organic. While wild fish may be free of pesticides and un-sustainable rearing practices, the fishing industry may not necessarily be environmentally sustainable.

The variation in standards, as well as the unknown level of actual compliance and the closeness of investigations when certifying are major problems in consistent organic certification. In 2010, new rules were proposed in the European Union to consistently define the organic aquaculture industry. Canada's General Standards Board's (CGSB) proposed updates to their standards were strongly opposed in 2010 because they allowed antibiotic and chemical treatments of fish, up to 30 percent non-organic feed, deadly and uncontrolled impacts on wild species and unrestrained disposal of fish feces into the ocean. These standards would have certified net pen systems as organic. At the other end of the scale, the extremely strict national legislation in Denmark has made it difficult for the existing organic trout industry to develop.

## Potential Alternatives to Non-organic Feed and Waste Removal

One major issue in organic aquaculture production is finding practical and sustainable alternatives to non-organic veterinary treatments, feeds, spat and waste disposal. Potential veterinary alternatives include homeopathic treatments and production-cycle limited allopathic or chemical treatments  Current requirements usually stipulate a reduction in unsustainable fishmeal, in favor of organic vegetable and fish by-product replacements. A recent study into organic fish feeds for salmon found that while organic feed provide some benefit to the environmental impact of the fishes' life cycles, the loss of fish meals and oils have a significant negative impact. Another study discovered that certain percentages of dietary protein could be safely replaced.

Not only do the fish have to be organically reared, organic fish feeds need to be developed. Research into ways of decreasing the amount on non-sustainable fishmeal in feed is currently focussing on replacement by organic vegetable proteins. Some organic fish feeds becoming available, and/or the option of integrated multi-species systems (e.g. growing plants using aquaponics, as well as larvae or other fish). For example, locating a shellfish bed next to a finfish farm to dispose of the waste and provide the shellfish with controlled nutrients.

## Certifying Bodies that Cover Organic Aquaculture

| Certification body | Countries of operation | No. of certified aquaculture farms | Accredited for grower groups | No. of certified groups | Aquaculture commodities within the scheme | Production (tonnes) |
|---|---|---|---|---|---|---|
| Agrior | Israel | 2 + 1 fish feed mill | no | NA | Tilapia, carp, red drum, sea bass, sea bream, Ulva and Ulea seaweed | 400 |
| AgriQuality Ltd. | New Zealand, Vanuatu, Cook Islands, Malaysia | yes | Example | | | |
| Bioland e.V. | Germany, Austria, Belgium, France, Italy, Netherlands, Switzerland | no | Example | | | |
| Debio | Norway | 3 | no | NA | salmon, trout, cod | trout 0.5 salmon 120 cod 600 |
| Instituto Biodinamico | Brazil, Argentina, Bolivia, Mexico, Paraguay, Uruguay | | yes | | | |
| Istituto per la Certificazione Etica e Ambientale | Italy, Lebanon, Turkey | | yes | | | |
| National Association Sustainable Agriculture Australia | Australia, Timor-Leste, Indonesia, Malaysia, Nepal, New Zealand, Papua New Guinea, Samoa, Sri Lanka, Solomon Islands | | yes | | | |
| Organic Agriculture Certification Thailand | Thailand | 1 (not under the IFOAM-accredited scheme) | Example | 0 | nile tilapia and butter fish | 8 000 litres (fish sauce) |

Table from IFOAM: Annex 6. Organic schemes

- United Kingdom The Soil Association

- Hungary Biokontrol Hungaria

- Naturland (Association for Organic Agriculture)

- Spain: Voluntary standards set by the Advisory Group CRAE do not cover organic aquaculture.

- New Zealand - BioGro

- Switzerland - Bio Suisse

- Nordic countries (Sweden, Norway) as well as Japan, Thailand and Australia - KRAV

## United States Organic Aquaculture Certification

In 2005, with the growing need for a certification process specifically designed for marine-based farming methods, the National Organic Standards Board and the National Organics Program created a working group called the Aquatic Animal Task Force in order to seek recommendations for the new certification process. The task force was meant to be broken into two divisions: wild fisheries and aquaculture, but the wild fisheries group never materialized.

In 2006, the Aquaculture Working Group delivered a report with suggestions for the production and handling of aquatic animals and plants. However, with the complexity and diversity of the marine systems, the group requested more time to explore bivalve mollusks (oysters, clams, mussels and scallops) in depth. The National Organic Standards Board approved the aquaculture standards in 2007 and reconsidered the aquatic animal feed and facilities until they synthesized the public commentary in 2008. In 2010, the NOSB approved the recommendations for the bivalve mollusks section.

Currently, the legal status of using the organic label for aquatic species, and the future of developing U.S. Department of Agriculture (USDA) certification standards for organic aquaculture products and aquatic species, are under review. It is anticipated that the first version of the rule for organic aquaculture will be announced in April or May 2016 with need for approval by the Office of Management and Budget. It is expected to see the final rule in play by late summer or fall of 2016 with organic aquaculture products likely available in store in 2017. The certification is said to include the following: shellfish, marine and recirculating system methods of aquaculture, as well as the controversial net-pen method.

The US currently allows the imports of organically-certified seafood from Europe, Canada and other countries around the world.

## Production

Organic aquaculture was responsible for an estimated US$46.1 billion internationally (2007). There were 0.4 million hectares of certified organic aquaculture in 2008 compared to 32.2 million hectares dedicated to Organic farming. The 2007 production was still only 0.1% of total aquaculture production

The market for organic aquaculture shows strong growth in Europe, especially France, Germany and the UK - for example, the market in France grew 220% from 2007 to 2008. There is a preference for organic food, where available. Organic seafood is now sold in discount supermarket chains throughout the EU. The top five producing countries are UK, Ireland, Hungary, Greece and France. 123 of the 225 global certified organic aquaculture farms operate in Europe and were responsible for 50,000 tonnes in 2008 (nearly half global production).

Organic seafood products are a niche market and users currently expect to pay premiums of 30-40%. Organic salmon is the top species and retails at 50%. Market demand is driving Danish rainbow trout farmers to switch to organic farming.

# Known Data on Organic Aquaculture by Country

## Asia

| Country | Organically managed area [ha] |
|---------|-------------------------------|
| Bangladesh | 2'000 |
| China | 415'000 |
| Ecuador | 6'382 |
| Indonesia[1] | 1'317 |
| Thailand | 33 |
| Total | 424'732 |

[1]Indonesian Shrimp farms are locally certified as organic but a recent study found them to be highly environmentally damaging.

## Europe

Current situation in Norway:

- Denmark: Rainbow Trout. Organic production ~400 tonnes (1% of total trout production)

- UK:

Cod and carp Trout Salmon

- Rainbow Trout (Denmark)

- Salmon (80% of organic aquaculture production in 2000 )

and shrimp (Europe)

- Carp (low volume production, poorly marketed - Europe)

## North America

- Shellfish: oyster, clam, mussel, scallop, geoduck seed (USA)

Organic production of crops and livestock in the United States is regulated by the Department of Agriculture's National Organic Program (NOP). While it does cover aquaponics, it did not properly cover aquaculture until the recent 2008 amendment, hampering the progress of organic aquaculture in the states.

## Australia

## New Zealand

The first certified organic aquaculture farm in New Zealand was a salmon farm which was the largest producer outside of Europe contributing to the European market. New Zealand green-lipped mussel Greenshell mussels - certified by Sealord (12), DOM ORGANICS Greenshell mussels, certified organic by Bio-Gro New Zealand Ltd. (BGNZ)

Salmon (14) 12 tonnes/year - Ormond Aquaculture Ltd certified (CERTNZ) organic freshwater aquaculture farm

Koura (freshwater crayfish) Still being developed - Ormond Aquaculture Ltd certified (CERTNZ) organic freshwater aquaculture farm

## Future Research and Development

Various methods and complementary processes are being investigated as alternatives for organic aquaculture, most notably Integrated Multi-Trophic Aquaculture(IMTA) and aquaponics (a land-based outgrowth of aquaculture in many places). Organic methods of farming various species are also topics of interest, particularly shrimps, salmon and Atlantic Cod

Projects such as ORAQUA are implementing scientific recommendations that support economic growth of Europe's organic aquaculture industry. The goals of this organization are as follows:

1. Reassess the relevance, measurability and applicability of Regulation EC 710/2009 for organic aquaculture against the basic organic principles;

2. Generate robust science based recommendations for potential updates of the EC regulation as regards aquaculture of fish species, molluscs, crustaceans and seaweed, based on comprehensive reviewing, research and assessment, in addition to integrating feedback from key stakeholders;

3. Produce executive dossiers on the main technical background behind the recommendations that will emerge from this project;

4. To underpin consumer demand for organic aquaculture products and development of organic aquaculture industry by integrating aspects of consumer perceptions, unique competitive qualities as well as production systems, business and market economics and regulatory framework;

5. To propose a model for continuous assessment and advice on the improvement of regulations of organic aquaculture in the future, taking account of new scientific insights and changing competitive market environments.

# Aquaponics

Aquaponics refers to any system that combines conventional aquaculture (raising aquatic animals such as snails, fish, crayfish or prawns in tanks) with hydroponics (cultivating plants in water) in a symbiotic environment. In normal aquaculture, excretions from the animals being raised can accumulate in the water, increasing toxicity. In an aquaponic system, water from an aquaculture system is fed to a hydroponic system where the by-products are broken down by Nitrifying bacteria into nitrates and nitrites, which are utilized by the plants as nutrients, and the water is then recirculated back to the aquaculture system.

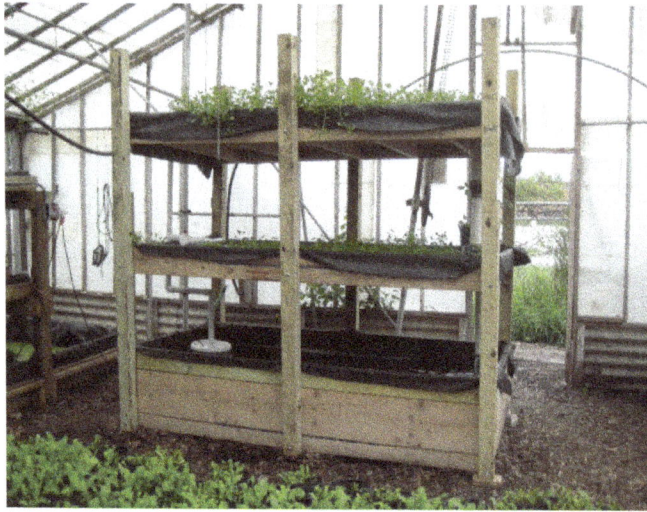

A small, portable aquaponics system. The term *aquaponics* is a portmanteau
of the terms *aquaculture* and *hydroponic* agriculture.

As existing hydroponic and aquaculture farming techniques form the basis for all aquaponics systems, the size, complexity, and types of foods grown in an aquaponics system can vary as much as any system found in either distinct farming discipline.

## History

Aquaponics has ancient roots, although there is some debate on its first occurrence:

- Aztec cultivated agricultural islands known as *chinampas* in a system considered by some to be the first form of aquaponics for agricultural use where plants were raised on stationary (and sometime movable) islands in lake shallows and waste materials dredged from the Chinampa canals and surrounding cities were used to manually irrigate the plants.

- South China, Thailand, and Indonesia who cultivated and farmed rice in paddy fields in combination with fish are cited as examples of early aquaponics systems. These polycultural farming systems existed in many Far Eastern countries and raised fish such as the oriental loach (泥鰍, ドジョウ), swamp eel (黄鱔, 田鰻), common carp (鯉魚, コイ) and crucian carp (鯽魚) as well as pond snails (田螺) in the paddies.

Floating aquaponics systems on polycultural fish ponds were installed in China in more recent years on a large scale growing rice, wheat and canna lily and other crops, with some installations exceeding 2.5 acres (10,000 m²).

The development of modern aquaponics is often attributed to the various works of the New Alchemy Institute and the works of Dr. Mark McMurtry et al. at the North Carolina State University. Inspired by the successes of the New Alchemy Institute, and the reciprocating aquaponics techniques developed by Dr. Mark McMurtry et al., other institutes soon followed suit. Starting in 1997, Dr. James Rakocy and his colleagues at the University of the Virgin Islands researched and developed the use of deep water culture hydroponic grow beds in a large-scale aquaponics system.

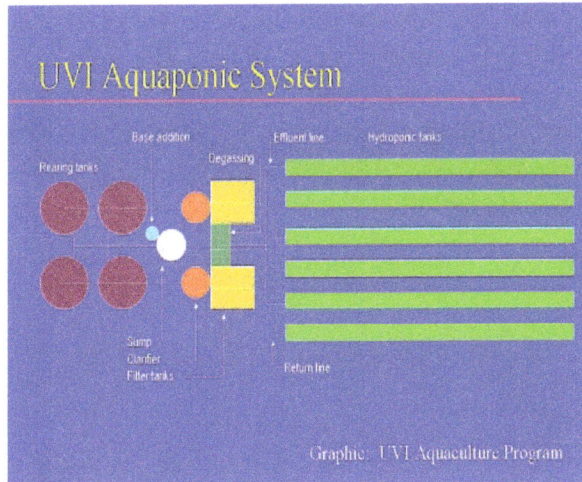

Diagram of the University of the Virgin Islands commercial aquaponics system
designed to yield 5 metric tons of Tilapia per year.

The first aquaponics research in Canada was a small system added onto existing aquaculture research at a research station in Lethbridge, Alberta. Canada saw a rise in aquaponics setups throughout the '90s, predominantly as commercial installations raising high-value crops such as trout and lettuce. A setup based on the deep water system developed at the University of Virgin Islands was built in a greenhouse at Brooks, Alberta where Dr. Nick Savidov and colleagues researched aquaponics from a background of plant science. The team made findings on rapid root growth in aquaponics systems and on closing the solid-waste loop, and found that owing to certain advantages in the system over traditional aquaculture, the system can run well at a low pH level, which is favoured by plants but not fish.

## Parts of an Aquaponic System

A commercial aquaponics system. An electric pump moves nutrient-rich water from the fish tank through a solids filter to remove particles the plants above cannot absorb. The water then provides nutrients for the plants and is cleansed before returning to the fish tank below.

Aquaponics consists of two main parts, with the aquaculture part for raising aquatic animals and the hydroponics part for growing plants. Aquatic effluents, resulting from uneaten feed or raising animals like fish, accumulate in water due to the closed-system recirculation of most aquaculture systems. The effluent-rich water becomes toxic to the aquatic animal in high concentrations but this contains nutrients essential for plant growth. Although consisting primarily of these two parts, aquaponics systems are usually grouped into several components or subsystems responsible for the effective removal of solid wastes, for adding bases to neutralize acids, or for maintaining water oxygenation. Typical components include:

- *Rearing tank*: the tanks for raising and feeding the fish;

- *Settling basin*: a unit for catching uneaten food and detached biofilms, and for settling out fine particulates;

- *Biofilter*: a place where the nitrification bacteria can grow and convert ammonia into nitrates, which are usable by the plants;

- *Hydroponics subsystem*: the portion of the system where plants are grown by absorbing excess nutrients from the water;

- *Sump*: the lowest point in the system where the water flows to and from which it is pumped back to the rearing tanks.

Depending on the sophistication and cost of the aquaponics system, the units for solids removal, biofiltration, and/or the hydroponics subsystem may be combined into one unit or subsystem, which prevents the water from flowing directly from the aquaculture part of the system to the hydroponics part.

## Live Components

An aquaponic system depends on different live components to work successfully. The three main live components are plants, fish (or other aquatic creatures) and bacteria. Some systems also include additional live components like worms.

## Plants

Many plants are suitable for aquaponic systems, though which ones work for a specific system depends on the maturity and stocking density of the fish. These factors influence the concentration of nutrients from the fish effluent, and how much of those nutrients are made available to the plant roots via bacteria.

Green leaf vegetables with low to medium nutrient requirements are well adapted to aquaponic systems, including chinese cabbage, lettuce, basil, spinach, chives, herbs, and watercress.

Other plants, such as tomatoes, cucumbers, and peppers, have higher nutrient requirements and will only do well in mature aquaponic systems that have high stocking densities of fish.

Plants that are common in salads have some of the greatest success in aquaponics, including cucumbers, shallots, tomatoes, lettuce, chiles, capsicum, red salad onions and snow peas.

A Deep Water Culture hydroponics system where plant grow directly into the effluent rich water without a soil medium. Plants can be spaced closer together because the roots do not need to expand outwards to support the weight of the plant.

Plant placed into a nutrient rich water channel in a Nutrient film technique (NFT) system

Some profitable plants for aquaponic systems include chinese cabbage, lettuce, basil, roses, tomatoes, okra, cantaloupe and bell peppers.

Other species of vegetables that grow well in an aquaponic system include watercress, basil, coriander, parsley, lemongrass, sage, beans, peas, kohlrabi, taro, radishes, strawberries, melons, onions, turnips, parsnips, sweet potato, cauliflower, cabbage, broccoli, and eggplant as well as the choys that are used for stir fries.

Fruiting plants like melons or tomatoes, and plants with higher nutrient needs need higher stocking densities of fish and more mature tanks to provide enough nutrients.

## Fish (or other Aquatic Creatures)

Filtered water from the hydroponics system drains into a catfish tank for re-circulation.

Freshwater fish are the most common aquatic animal raised using aquaponics, although freshwater crayfish and prawns are also sometimes used. There is a branch of aquaponics using saltwater fish, called saltwater aquaponics. There are many species of warmwater and coldwater fish that adapt well to aquaculture systems.

In practice, tilapia are the most popular fish for home and commercial projects that are intended to raise edible fish because it is a warmwater fish species that can tolerate crowding and changing water conditions. Barramundi, silver perch, eel-tailed catfish or tandanus catfish, jade perch and Murray cod are also used. For temperate climates when there isn't ability or desire to maintain water temperature, bluegill and catfish are suitable fish species for home systems.

Koi and goldfish may also be used, if the fish in the system need not be edible.

Other suitable fish include channel catfish, rainbow trout, perch, common carp, Arctic char, largemouth bass and striped bass.

## Bacteria

Nitrification, the aerobic conversion of ammonia into nitrates, is one of the most important functions in an aquaponics system as it reduces the toxicity of the water for fish, and allows the resulting nitrate compounds to be removed by the plants for nourishment. Ammonia is steadily released into the water through the excreta and gills of fish as a product of their metabolism, but must be filtered out of the water since higher concentrations of ammonia (commonly between 0.5 and 1 ppm) can kill fish. Although plants can absorb ammonia from the water to some degree, nitrates are assimilated more easily, thereby efficiently reducing the toxicity of the water for fish. Ammonia can be converted into other nitrogenous compounds through combined healthy populations of:

- *Nitrosomonas*: bacteria that convert ammonia into nitrites, and

- *Nitrobacter*: bacteria that convert nitrites into nitrates.

## Hydroponic Subsystem

Plants are grown as in hydroponics systems, with their roots immersed in the nutrient-rich effluent water. This enables them to filter out the ammonia that is toxic to the aquatic animals, or its metabolites. After the water has passed through the hydroponic subsystem, it is cleaned and oxygenated, and can return to the aquaculture vessels. This cycle is continuous. Common aquaponic applications of hydroponic systems include:

- *Deep-water raft aquaponics*: styrofoam rafts floating in a relatively deep aquaculture basin in troughs.

- *Recirculating aquaponics*: solid media such as gravel or clay beads, held in a container that is flooded with water from the aquaculture. This type of aquaponics is also known as *closed-loop aquaponics*.

- *Reciprocating aquaponics*: solid media in a container that is alternately flooded and drained utilizing different types of siphon drains. This type of aquaponics is also known as *flood-and-drain aquaponics* or *ebb-and-flow aquaponics*.

- Other systems use towers that are trickle-fed from the top, nutrient film technique channels, horizontal PVC pipes with holes for the pots, plastic barrels cut in half with gravel or rafts in them. Each approach has its own benefits.

Since plants at different growth stages require different amounts of minerals and nutrients, plant harvesting is staggered with seedlings growing at the same time as mature plants. This ensures stable nutrient content in the water because of continuous symbiotic cleansing of toxins from the water.

## Biofilter

In an aquaponics system, the bacteria responsible for the conversion of ammonia to usable nitrates for plants form a biofilm on all solid surfaces throughout the system that are in constant contact with the water. The submerged roots of the vegetables combined have a large surface area where many bacteria can accumulate. Together with the concentrations of ammonia and nitrites in the water, the surface area determines the speed with which nitrification takes place. Care for these bacterial colonies is important as to regulate the full assimilation of ammonia and nitrite. This is why most aquaponics systems include a biofiltering unit, which helps facilitate growth of these microorganisms. Typically, after a system has stabilized ammonia levels range from 0.25 to 2.0 ppm; nitrite levels range from 0.25 to 1 ppm, and nitrate levels range from 2 to 150 ppm. During system startup, spikes may occur in the levels of ammonia (up to 6.0 ppm) and nitrite (up to 15 ppm), with nitrate levels peaking later in the startup phase. Since the nitrification process acidifies the water, non-sodium bases such as potassium hydroxide or calcium hydroxide can be added for neutralizing the water's pH if insufficient quantities are naturally present in the water to provide a buffer against acidification. In addition, selected minerals or nutrients such as iron can be added in addition to the fish waste that serves as the main source of nutrients to plants.

A good way to deal with solids buildup in aquaponics is the use of worms, which liquefy the solid organic matter so that it can be utilized by the plants and/or other animals in the system.

## Operation

The five main inputs to the system are water, oxygen, light, feed given to the aquatic animals, and electricity to pump, filter, and oxygenate the water. Spawn or fry may be added to replace grown fish that are taken out from the system to retain a stable system. In terms of outputs, an aquaponics system may continually yield plants such as vegetables grown in hydroponics, and edible aquatic species raised in an aquaculture. Typical build ratios are .5 to 1 square foot of grow space for every 1 U.S. gal (3.8 L) of aquaculture water in the system. 1 U.S. gal (3.8 L) of water can support between .5 lb (0.23 kg) and 1 lb (0.45 kg) of fish stock depending on aeration and filtration.

Ten primary guiding principles for creating successful aquaponics systems were issued by Dr. James Rakocy, the director of the aquaponics research team at the University of the Virgin Islands, based on extensive research done as part of the *Agricultural Experiment Station* aquaculture program.

- Use a feeding rate ratio for design calculations

- Keep feed input relatively constant

- Supplement with calcium, potassium and iron

- Ensure good aeration

- Remove solids

- Be careful with aggregates

- Oversize pipes

- Use biological pest control

- Ensure adequate biofiltration

- Control pH

## Feed Source

As in most aquaculture based systems, stock feed often consists of fish meal derived from lower-value species. Ongoing depletion of wild fish stocks makes this practice unsustainable. Organic fish feeds may prove to be a viable alternative that relieves this concern. Other alternatives include growing duckweed with an aquaponics system that feeds the same fish grown on the system, excess worms grown from vermiculture composting, using prepared kitchen scraps, as well as growing black soldier fly larvae to feed to the fish using composting grub growers.

## Water Usage

Aquaponic systems do not typically discharge or exchange water under normal operation, but instead recirculate and reuse water very effectively. The system relies on the relationship between the animals and the plants to maintain a stable aquatic environment that experience a minimum of fluctuation in ambient nutrient and oxygen levels. Water is added only to replace water loss from absorption and transpiration by plants, evaporation into the air from surface water, overflow from the system from rainfall, and removal of biomass such as settled solid wastes from the system. As a result, aquaponics uses approximately 2% of the water that a conventionally irrigated farm requires for the same vegetable production. This allows for aquaponic production of both crops and fish in areas where water or fertile land is scarce. Aquaponic systems can also be used to replicate controlled wetland conditions. Constructed wetlands can be useful for biofiltration and treatment of typical household sewage. The nutrient-filled overflow water can be accumulated in catchment tanks, and reused to accelerate growth of crops planted in soil, or it may be pumped back into the aquaponic system to top up the water level.

## Energy Usage

Aquaponic installations rely in varying degrees on man-made energy, technological solutions, and environmental control to achieve recirculation and water/ambient temperatures. However, if a system is designed with energy conservation in mind, using alternative energy and a reduced number of pumps by letting the water flow downwards as much as possible, it can be highly energy efficient. While careful design can minimize the risk, aquaponics systems can have multiple 'single points of failure' where problems such as an electrical failure or a pipe blockage can lead to a complete loss of fish stock.

## Current Examples

The Caribbean island of Barbados created an initiative to start aquaponics systems at home, with revenue generated by selling produce to tourists in an effort to reduce growing dependence on imported food.

Dakota College at Bottineau in Bottineau, North Dakota has an aquaponics program that gives students the ability to a obtain a certificate or an AAS degree in aquaponics.

Vegetable production part of the low-cost Backyard Aquaponics System developed at Bangladesh Agricultural University

In Bangladesh, the world's most densely populated country, most farmers use agrochemicals to enhance food production and storage life, though the country lacks oversight on safe levels of chemicals in foods for human consumption. To combat this issue, a team led by Professor Dr. M.A. Salam at the Department of Aquaculture of Bangladesh Agricultural University, Mymensingh has created plans for a low-cost aquaponics system to provide chemical-free produce and fish for people living in adverse climatic conditions such as the salinity-prone southern area and the flood-prone haor area in the eastern region. Dr. Salam's work innovates a form of subsistence farming for micro-production goals at the community and personal levels whereas design work by Chowdhury and Graff was aimed exclusively at the commercial level, the latter of the two approaches take advantage of economies of scale.

With more than a third of Palestinian agricultural lands in the Gaza Strip turned into a buffer zone by Israel, an aquaponic gardening system is developed appropriate for use on rooftops in Gaza City.

There has been a shift towards community integration of aquaponics, such as the nonprofit foundation Growing Power that offers Milwaukee youth job opportunities and training while growing food for their community. The model has spawned several satellite projects in other cities, such as New Orleans where the Vietnamese fisherman community has suffered from the Deepwater Horizon oil spill, and in the South Bronx in New York City.

Whispering Roots is a non-profit organization in Omaha, Nebraska that provides fresh, locally grown, healthy food for socially and economically disadvantaged communities by using aquaponics, hydroponics and urban farming.

In addition, aquaponic gardeners from all around the world are gathering in online community sites and forums to share their experiences and promote the development of this form of gardening as well as creating extensive resources on how to build home systems.

Recently, aquaponics has been moving towards indoor production systems. In cities like Chicago, entrepreneurs are utilizing vertical designs to grow food year round. These systems can be used to grow food year round with minimal to no waste.

There are various modular systems made for the public that utilize aquaponic systems to produce organic vegetables and herbs, and provide indoor decor at the same time. These systems can serve as a source of herbs and vegetables indoors. Universities are promoting research on these modular systems as they get more popular among city dwellers.

# Copper Alloys in Aquaculture

Copper alloys are important netting materials in aquaculture (the farming of aquatic organisms including fish farming). Various other materials including nylon, polyester, polypropylene, polyethylene, plastic-coated welded wire, rubber, patented twine products (Spectra, Dyneema), and galvanized steel are also used for netting in aquaculture fish enclosures around the world. All of these materials are selected for a variety of reasons, including design feasibility, material strength, cost, and corrosion resistance.

What sets copper alloys apart from the other materials used in fish farming is that copper alloys are antimicrobial, that is, they destroy bacteria, viruses, fungi, algae, and other microbes.

A copper alloy pen that has been deployed on a fish farm at depth of 14 feet for one year shows no signs of biofouling.

In the marine environment, the antimicrobial/algaecidal properties of copper alloys prevent biofouling, which can briefly be described as the undesirable accumulation, adhesion, and growth of microorganisms, plants, algae, tube worms, barnacles, mollusks, and other organisms on man-made marine structures. By inhibiting microbial growth, copper alloy aquaculture pens avoid the need for costly net changes that are necessary with other materials. The resistance of organism growth on copper alloy nets also provides a cleaner and healthier environment for farmed fish to grow and thrive.

In addition to their antifouling benefits, copper alloys have strong structural and corrosion-resistant properties in marine environments.

It is the combination of all of these properties – antifouling, high strength, and corrosion resistance – that has made copper alloys a desirable material for such marine applications as condenser tubing, water intake screens, ship hulls, offshore structure, and sheathing. In the past 25 years or so, the benefits of copper alloys have caught the attention of the marine aquaculture industry. The industry is now actively deploying copper alloy netting and structural materials in commercial large-scale fish farming operations around the world.

## Importance of Aquaculture

Much has been written about the degradation and depletion of natural fish stocks in rivers, estuaries, and the oceans. Because industrial fishing has become extremely efficient, ocean stocks of large fish, such as tuna, cod, and halibut have declined by 90% in the past 50 years.

Aquaculture, an industry that has emerged only in recent decades, has become one of the fastest growing sectors of the world food economy. Aquaculture already supplies more than half of the world's demand for fish. This percentage is predicted to increase dramatically over the next few decades.

## The Problem of Biofouling

Copper alloy mesh installed at an Atlantic salmon fish farm in Tasmania. Foreground: the chain link copper alloy mesh resting on a dock. Distant background: copper alloy mesh pens are installed on the fish farm.

Biofouling is one of the biggest problems in aquaculture. Biofouling occurs on non-copper materials in the marine environment, including fish pen surfaces and nettings. For example, it was noted that the open area of a mesh immersed for only seven days in a Tasmanian aquaculture operation decreased by 37% as a result of biofouling.

The biofouling process begins when algae spores, marine invertebrate larvae, and other organic material adhere to surfaces submerged in marine environments (e.g., fish nets in aquaculture). Bacteria then encourage the attachment of secondary unwanted colonizers.

Biofouling has strong negative impacts on aquaculture operations. Water flow and dissolved oxygen are inhibited due to clogged nets in fish pens. The end result is often diseased fish from infections, such as netpen liver disease, amoebic gill disease, and parasites. Other negative impacts include increased fish mortalities, decreased fish growth rates, premature fish harvesting, reduced fish product values and profitability, and an adversely impacted environment near fish farms.

Biofouling adds enormous weight to submerged fish netting. Two hundredfold increases in weight have been reported. This translates, for example, to two thousand pounds of unwanted organisms adhered to what was once a clean 10-pound fish pen net. In South Australia, biofouling weighing 6.5 tonnes (approximately 13,000 pounds) was observed on a fish pen net. This extra burden often results in net breakage and additional maintenance costs.

To combat parasites from biofouling in finfish aquaculture, treatment protocols such as cypermethrin, azamethiphos, and emamectin benzoate may be administered, but these have been found to have detrimental environmental effects, for example, in lobster operations.

To treat diseases in fish raised in biofouled nets, fish stocks are administered antibiotics. The antibiotics can have unwanted long-term health effects on consumers and on coastal environments near aquaculture operations. To combat biofouling, operators often implement costly maintenance measures, such as frequent net changing, cleaning/removal of unwanted organisms from nets, net repairs, and chemical treatment including antimicrobial coatings on nylon nets. The cost of antifouling a single salmon net can be several thousand British pounds. In some sectors of the

European aquaculture industry, cleaning biofouled fish and shellfish pens can cost 5–20% of its market value. Heavy fouling can reduce the saleable product in nets by 60–90%.

Antifouling coatings are often used on nylon nets because the process is more economical than manual cleaning. When nylon nets are coated with antifouling compounds, the coatings repel biofouling for a period of time, usually between several weeks to several months. However, the nets eventually succumb to biofouling. Antifouling coatings containing cuprous oxide algaecide/biocide are the coatings technology used almost exclusively in the fish farming industry today. The treatments usually flake off within a few weeks to six to eight months.

Biofouled nets are replaced after several months of service, depending on environmental conditions, in a complicated, costly, and labor-intensive operation that involves divers and specialized personnel. During this process, live fish in nets must be transferred to clean pens, which causes undue stress and asphyxiation that results in some loss of fish. Biofouled nets that can be reused are washed on land via manual brushing and scrubbing or high-pressure water hosing. They are then dried and re-impregnated with antifouling coatings.

A line of net cleaners is available for in-situ washings where permitted. But, even where not permitted by environmental, fisheries, maritime, and sanitary authorities, should the lack of dissolved oxygen in submerged pens create an emergency condition that endangers the health of fish, divers may be deployed with special in situ cleaning machinery to scrub biofouled nets.

The aquaculture industry is addressing the negative environmental impacts from its operations. As the industry evolves, a cleaner, more sustainable aquaculture industry is expected to emerge, one that may increasingly rely on materials with anti-fouling, anti-corro-sive, and strong structural properties, such as copper alloys.

## Antifouling Properties of Copper Alloys

In the aquaculture industry, sound animal husbandry translates to keeping fish clean, well fed, healthy, and not overcrowded. One solution to keeping farmed fish healthy is to contain them in antifouling copper alloy nets and structures.

There is no biofouling on a copper alloy mesh after 4 months immersed in the waters of the North Atlantic (foreground), whereas hydroids have grown on high-density polyethylene tubing (background).

Researchers have attributed copper's resistance to biofouling, even in temperate waters, to two possible mechanisms: 1) a retarding sequence of colonization through release of antimicrobial copper ions, thereby preventing the attachment of microbial layers to marine surfaces; and, 2) separating layers that contain corrosive products and the spores of juveniles or macro-encrusting organisms.

The most important requirement for optimum biofouling resistance is that the copper alloys should be freely exposed or electrically insulated from less noble alloys and from cathodic protection. Galvanic coupling to less noble alloys and cathodic protection prevent copper ion releases from surface films and therefore reduce biofouling resistance.

As temperatures increase and water velocities decrease in marine waters, biofouling rates dramatically rise. However, copper's resistance to biofouling is observed even in temperate waters. Studies in La Herradura Bay, Coquimbo, Chile, where biofouling conditions are extreme, demonstrated that a copper alloy (90% copper, 10% nickel) avoided macro-encrusting organisms.

## Corrosion Behavior of Copper Alloys

Copper alloys used in sea water service have low general corrosion rates but also have a high resistance to many localized forms of corrosion. A technical discussion regarding various types of corrosion, application considerations (e.g., depth of installations, effect of polluted waters, sea conditions), and the corrosion characteristics of several copper alloys used in aquaculture netting is available (i.e., copper-nickel, copper-zinc, and copper-silicon).

## Early Examples of Copper Sheathing

Prior to the late 1700s, hulls were made almost entirely of wood, often white oak. Sacrificial planking was the common mode of hull protection. This technique included wrapping a protective 1/2-inch thick layer of wood, often pine, on the hull to decrease the risk of damage. This layer was replaced regularly when infested with marine borers. Copper sheathing for bio-resistant ship hulls was developed in the late 18th century. In 1761, the hull of the British Royal Navy's HMS Alarm frigate was fully sheathed in copper to prevent attack by Teredo worms in tropical waters. The copper reduced biofouling of the hull, which enabled ships to move faster than those that did not have copper sheathed hulls.

## Environmental Performance of Copper Alloy Mesh

Many complicated factors influence the environmental performance of copper alloys in aquaculture operations. A technical description of antibiofouling mechanisms, fish health and welfare, fish losses due to escapes and predator attacks, and reduced life cycle environmental impacts is summarized in this reference.

## Types of Copper Alloys

Copper–zinc brass alloys are currently (2011) being deployed in commercial-scale aquaculture operations in Asia, South America and the US (Hawaii). Extensive research, including demonstrations and trials, are currently being implemented on two other copper alloys: copper-nickel and copper-silicon. Each of these alloy types has an inherent ability to reduce biofouling, pen waste, disease, and the need for antibiotics while simultaneously maintaining water circulation and oxygen requirements. Other types of copper alloys are also being considered for research and development in aquaculture operations.

The University of New Hampshire is in the midst of conducting experiments under the auspices of the International Copper Association (ICA) to evaluate the structural, hydrodynamic, and antifouling response of copper alloy nets. Factors to be determined from these experiments, such as drag, pen dynamic loads, material loss, and biological growth – well documented for nylon netting but not fully understood for copper-nickel alloy nets – will help to design fish pen enclosures made from these alloys. The East China Sea Fisheries Research Institute, in Shanghai, China, is also conducting experimental investigations on copper alloys for ICA.

Section of a fish net on a salmon farm near Puerto Montt, Chile. The copper alloy woven mesh inside the frame has resisted biofouling whereas PVC (i.e., the frame around the mesh) is heavily fouled.

## Copper–zinc Alloys

The Mitsubishi-Shindoh Co., Ltd., has developed a proprietary copper-zinc brass alloy, called UR30, specifically designed for aquaculture operations. The alloy, which is composed of 64% copper, 35.1% zinc, 0.6% tin, and 0.3% nickel, resists mechanical abrasion when formed into wires and fabricated into chain link, woven, or other types of flexible mesh. Corrosion rates depend on the depth of submersion and seawater conditions. The average reported corrosion rate reported for the alloy is < 5 μm/yr based on two- and five-year exposure trials in seawater.

The Ashimori Industry Company, Ltd., has installed approximately 300 flexible pens with woven chain link UR30 meshes in Japan to raise Seriola (i.e., yellowtail, amberjack, kingfish, hamachi). The company has installed another 32 brass pens to raise Atlantic salmon at the Van Diemen Aquaculture operations in Tasmania, Australia. In Chile, EcoSea Farming S.A. has installed a total of 62 woven chain link brass mesh pens to raise trout and Atlantic salmon. In Panama, China, Korea, Turkey, and the US, demonstrations and trials are underway using flexible pens with woven chain link UR30 and other mesh forms and a range of copper alloys.

To date, in over 10 years of aquaculture experience, chain link mesh fabricated by these brass alloys have not suffered from dezincification, stress corrosion cracking, or erosion corrosion.

## Copper–nickel Alloys

Copper–nickel alloys were developed specifically for seawater applications over five decades ago. Today, these alloys are being investigated for their potential use in aquaculture.

Copper–nickel alloys for marine applications are usually 90% copper, 10% nickel, and small amounts of manganese and iron to enhance corrosion resistance. The seawater corrosion resistance of copper–nickel alloys results in a thin, adherent, protective surface film which forms naturally and quickly on the metal upon exposure to clean seawater.

The rate of corrosion protective formation is temperature dependent. For example, at 27 °C (i.e., a common inlet temperature in the Middle East), rapid film formation and good corrosion protection can be expected within a few hours. At 16 °C, it could take 2–3 months for the protection to mature. But once a good surface film forms, corrosion rates decrease, normally to 0.02–0.002 mm/yr, as protective layers develop over a period of years. These alloys have good resistance to chloride pitting and crevice corrosion and are not susceptible to chloride stress corrosion.

## Copper–silicon Alloys

Copper–silicon has a long history of use as screws, nuts, bolts, washers, pins, lag bolts, and staples in wooden sailing vessels in marine environments. The alloys are often composed of copper, silicon, and manganese. The inclusion of silicon strengthens the metal.

As with the copper–nickel alloys, corrosion resistance of copper–silicon is due to protective films that form on the surface over a period of time. General corrosion rates of 0.025–0.050mm have been observed in quiet waters. This rate decreases towards the lower end of the range over long-term exposures (e.g., 400–600 days). There is generally no pitting with the silicon-bronzes. Also there is good resistance to erosion corrosion up to moderate flow rates. Because copper–silicon is weldable, rigid pens can be constructed with this material. Also, because welded copper–silicon mesh is lighter than copper-zinc chain link, aquaculture enclosures made with copper–silicon may be lighter in weight and therefore a potentially less expensive alternative.

Luvata Appleton, LLC, is researching and developing a line of copper alloy woven and welded meshes, including a patent-pending copper silicon alloy, that are marketed under the trade name Seawire. Copper-silicon alloy meshes have been developed by the firm to raise various marine organisms in test trials that are now in various stages of evaluation. These include raising cobia in Panama, lobsters in the US state of Maine, and crabs in the Chesapeake Bay. The company is working with various universities to study its material, including the University of Arizona to study shrimp, the University of New Hampshire to study cod, and Oregon State University to study oysters.

## References

- Michael B. Timmons and James B. Ebeling (2013). Recirculating Aquaculture (3rd ed.). Ithaca Publishing Company Publishers. p. 3. ISBN 978-0971264656.

- Tietenberg, T.H. (2006), Environmental and Natural Resource Economics: A Contemporary Approach, p. 28, Pearson/Addison Wesley. ISBN 0-321-30504-3

- Milne, P.H., (1970), Fish Farming: A guide to the design and construction of net enclosures, Marine Research, Vol. 1, pp. 1–31 ISBN 0-11-490463-4

- Li, S. (1994), Fish culture in cages and pens: Freshwater Fish Culture in China: Principles and Practice, pp. 305–346, Elsevier, Amsterdam ISBN 0-444-88882-9

- "USDA organic aquaculture label could hit grocery shelves in 2017, government says". Undercurrent News. Retrieved 2016-04-23.

- "COMMENTS and PROPOSED REVISIONS by the Aquaculture Working Group Pertaining to the Recommendations of the USDA National Organic Standards Board for Organic Aquaculture Standards" (PDF). USDA. USDA. October 2010. Retrieved April 23, 2016.

- Jensen, Thomas I. "Bremen: International Aquaculture Workshop brings together participants from all over Europe". www.eurofish.dk. Retrieved 2016-04-22.

- Losordo, T.; Massar, M.; Rakocy, J (September 1998). "Recirculating Aquaculture Tank Production Systems: an overview of critical conditions" (PDF). Retrieved August 25, 2015.

- Malone, Ron (October 2013). "Recirculating Aquaculture Tank Production Systems: A Review of Current Design Practices" (PDF). North Carolina State University. p. 5. Retrieved October 3, 2015.

- Yanong, R. "Fish Health Management Considerations in Recirculating Aquaculture Systems - Part 1: Introduction and General Principles" (PDF). Retrieved August 25, 2015.

- Rawlinson, P.; Forster, A. (2000). "The Economics of Recirculation Aquaculture" (PDF). Oregon State University. Retrieved October 3, 2015.

- Jenner, Andrew (February 24, 2010). "Recirculating aquaculture systems: The future of fish farming?". Christian Science Monitor. Retrieved August 25, 2015.

- Hall, Antar (December 1, 1999). A Comparative Analysis of Three Biofilter Types Treating Wastewater Produced in Recirculating Aquaculture Systems (PDF) (Master of Science). Retrieved September 16, 2015.

# 6

# Fisheries: An Integrated Field

Fisheries can be defined as any region that cultivates and farms fish and other marine animals for personal or commercial use. Fisheries provide livelihood for millions of the population, especially in the third world. This chapter is an overview of the subject matter incorporating all the major aspects of fisheries.

## Fishery

Salmon spawn in a salmon fishery within the Becharof Wilderness in Southwest Alaska.

Generally, a fishery is an entity engaged in raising or harvesting fish which is determined by some authority to be a fishery. According to the FAO, a fishery is typically defined in terms of the "people involved, species or type of fish, area of water or seabed, method of fishing, class of boats, purpose of the activities or a combination of the foregoing features". The definition often includes a combination of fish and fishers in a region, the latter fishing for similar species with similar gear types.

A fishery may involve the capture of wild fish or raising fish through fish farming or aquaculture. Directly or indirectly, the livelihood of over 500 million people in developing countries depends on fisheries and aquaculture. Overfishing, including the taking of fish beyond sustainable levels, is reducing fish stocks and employment in many world regions. A report by Prince Charles' International Sustainability Unit, the New York-based Environmental Defense Fund and 50in10 published in July 2014 estimated global fisheries were adding $270 billion a year to global GDP, but by full implementation of sustainable fishing, that figure could rise by as much as $50 billion.

### The Term Fish

- In biology – the term *fish* is most strictly used to describe any animal with a backbone that has gills throughout life and has limbs, if any, in the shape of fins. Many types of aquatic animals commonly referred to as *fish* are not fish in this strict sense; examples include shellfish, cuttlefish, starfish, crayfish and jellyfish. In earlier times, even biologists did not make a distinction — sixteenth century natural historians classified also seals, whales, amphibians, crocodiles, even hippopotamuses, as well as a host of marine invertebrates, as fish.

- In fisheries – the term *fish* is used as a collective term, and includes mollusks, crustaceans and any aquatic animal which is harvested.

- True fish – The strict biological definition of a fish, above, is sometimes called a true fish. True fish are also referred to as *finfish* or *fin fish* to distinguish them from other aquatic life harvested in fisheries or aquaculture.

## Types

Fishermen in Sesimbra, Portugal

Fisheries are harvested for their value (commercial, recreational or subsistence). They can be saltwater or freshwater, wild or farmed. Examples are the salmon fishery of Alaska, the cod fishery off the Lofoten islands, the tuna fishery of the Eastern Pacific, or the shrimp farm fisheries in China. Capture fisheries can be broadly classified as industrial scale, small-scale or artisanal, and recreational.

Close to 90% of the world's fishery catches come from oceans and seas, as opposed to inland waters. These marine catches have remained relatively stable since the mid-nineties (between 80 and 86 million tonnes). Most marine fisheries are based near the coast. This is not only because harvesting from relatively shallow waters is easier than in the open ocean, but also because fish are much more abundant near the coastal shelf, due to the abundance of nutrients available there from coastal upwelling and land runoff. However, productive wild fisheries also exist in open oceans, particularly by seamounts, and inland in lakes and rivers.

Most fisheries are wild fisheries, but farmed fisheries are increasing. Farming can occur in coastal areas, such as with oyster farms, but more typically occur inland, in lakes, ponds, tanks and other enclosures.

There are species fisheries worldwide for finfish, mollusks, crustaceans and echinoderms, and by extension, aquatic plants such as kelp. However, a very small number of species support the majority of the world's fisheries. Some of these species are herring, cod, anchovy, tuna, flounder, mullet, squid, shrimp, salmon, crab, lobster, oyster and scallops. All except these last four provided a worldwide catch of well over a million tonnes in 1999, with herring and sardines together providing a harvest of over 22 million metric tons in 1999. Many other species are harvested in smaller numbers.

# Fisheries Science

The 78-metre Danish fisheries research vessel *Dana*

Fisheries science is the academic discipline of managing and understanding fisheries. It is a multidisciplinary science, which draws on the disciplines of limnology, oceanography, freshwater biology, marine biology, conservation, ecology, population dynamics, economics and management to attempt to provide an integrated picture of fisheries. In some cases new disciplines have emerged, as in the case of bioeconomics and fisheries law.

Fisheries science is typically taught in a university setting, and can be the focus of an undergraduate, master's or Ph.D. program. Some universities offer fully integrated programs in fisheries science.

## Fisheries Research

Fisheries research vessels (FRVs) require platforms which are capable of towing different types of fishing nets, collecting plankton or water samples from a range of depths, and carrying acoustic fish-finding equipment. Fisheries research vessels are often designed and built along the same lines as a large fishing vessel, but with space given over to laboratories and equipment storage, as opposed to storage of the catch.

## Notable Contributors

Members of this list meet one or more of the following criteria: 1) Author of widely cited peer-reviewed articles on fisheries, 2) Author of major reference work in fisheries, 3) Founder of major fisheries journal, museum or other related organisation 4) Person most notable for other reasons who has also worked in fisheries science.

Ransom A. Myers

Daniel Pauly

Ray Hilborn

| Contributor | Nationality | Born | Died | Contribution |
|---|---|---|---|---|
| Baird, Spencer F | American | 1823 | 1887 | Founding scientist of the United States Fish Commission |
| Baranov, Fedor I | Russian | 1886 | 1965 | Baranov has been called the grandfather of fisheries population dynamics. The Baranov catch equation of 1918 is perhaps the most used equation in fisheries modelling. |
| Beverton, Ray | English | 1922 | 1985 | Fisheries biologist known for the Beverton–Holt model (with Sidney Holt), credited with being one of the founders of fisheries science |
| Christensen, Villy | Danish | | - | Fisheries scientist and ecosystem modeller, known for his work on the development of Ecopath |
| Cobb, John N | American | 1868 | 1930 | Founder of the first college of fisheries in the United States, the University of Washington College of Fisheries, in 1919 |
| Cooke, Steven J | Canadian | 1974 | | Academic known for contributions to recreational fisheries science, inland fisheries and Conservation Physiology |
| Cushing, David | English | 1920 | 2008 | Fisheries biologist, who is credited with the development of the match/mismatch hypothesis |
| Everhart, W Harry | American | 1918 | 1994 | Fisheries scientist, educator, administrator and author of several widely used fisheries texts |
| Froese, Rainer | German | 1950 | - | Known for his work on the development and coordination of FishBase |
| Green, Seth | American | 1817 | 1888 | Pioneer in fish farming who established the first fish hatchery in the United States |
| Halver, John | American | 1922 | 2012 | His pioneering work on the nutritional needs of fish led to modern methods of fish farming and fish feed production. He has been called the father of fish nutrition. |
| Hempel, Gotthilf | German | 1929 | - | Marine biologist and oceanographer, and co-founder of the Alfred Wegener Institute for Polar and Marine Research |
| Herwig, Walther | German | 1838 | 1912 | Lawyer and promoter of high seas fishing and research |
| Hilborn, Ray | Canadian | 1947 | - | Fisheries biologist with strong contributions in fisheries management |
| Hjort, Johan | Norwegian | 1869 | 1948 | Fisheries biologist, marine zoologist and oceanographer |
| Hofer, Bruno | German | 1861 | 1916 | Fishery scientist credited with being the founder of fish pathology |
| Holt, Sidney | English | 1926 | - | Fisheries biologist known for the Beverton–Holt model (with Ray Beverton), credited with being one of the founders of fisheries science |
| Kils, Uwe | German | | - | Marine biologist specializing in planktology. Inventor of the ecoSCOPE |
| Lackey, Robert T | Canadian | 1944 | - | Fisheries scientist and political scientist known for his work involving the role of science in policy making |
| Margolis, Leo | Canadian | 1927 | 1997 | Parasitologist and head of the Pacific Biological Station in Nanaimo, British Columbia |
| McKay, R J | Australian | | | Biologist and a specialist in translocated freshwater fishes |
| Myers, Ransom A | Canadian | 1952 | 2007 | Marine biologist and conservationist |
| Pauly, Daniel | French | 1946 | | Prominent fisheries scientist, known for his work studying human impacts on global fisheries |

| Contributor | Nationality | Born | Died | Contribution |
|---|---|---|---|---|
| Pitcher, Tony J | | | - | Known for work on the impacts of fishing, management appraisals and the shoaling behavior of fish |
| Rice, Michael A | American | 1955 | - | Known for work on molluscan fisheries |
| Ricker, Bill | Canadian | 1908 | 2001 | Fisheries biologist, known for the Ricker model, credited with being one of the founders of fisheries science |
| Ricketts, Ed | American | 1897 | 1948 | A colourful marine biologist and philosopher who introduced ecology to fisheries science. |
| Roberts, Callum | | | - | Marine conservation biologist, known for his work on the role marine reserves play in protecting marine ecosystems |
| Rosenthal, Harald | German | 1937 | - | Hydrobiologist known for his work in fish farming and ecology |
| Safina, Carl | American | 1955 | - | Author of several writings on marine ecology and the ocean |
| Sars, Georg Ossian | Norwegian | 1837 | 1927 | Marine biologist credited with the discovery of a number of new species and known for his analysis of cod fisheries |
| Schaefer, Milner Baily | American | 1912 | 1970 | Notable for work on the population dynamics of fisheries |
| Schweder, Tore | Norwegian | 1943 | - | Statistician whose work includes the assessment of marine resources |
| Sumaila, Ussif Rashid | Nigerian | | - | Notable for his analysis of the economic aspects of fisheries |
| Utter, Fred M | American | 1931 | - | Notable as the founding father of the field of fishery genetics and his influence on marine conservation |
| von Bertalanffy, Ludwig | Austrian | 1901 | 1972 | Biologist and founder of general systems theory |
| Walters, Carl | American | | - | Biologist known for his work involving fisheries stock assessments, the adaptive management concept, and ecosystem modeling |

# Journals

Fisheries scientists sorting a catch of small fish and Norway lobster

Some journals about fisheries are

- *Journal of Fisheries*

- *Fishery Bulletin*

- *Fisheries Oceanography*

- *Journal of the Fisheries Research Board*

- *Canadian Journal of Fisheries and Aquatic Sciences*

- *Transactions of the American Fisheries Society*

- *Fisheries Management and Ecology*

- *Fish and Fisheries*

- *Journal of Fish Biology*

- *Journal of Northwest Atlantic Fishery Science*

- *Journal of Fisheries and Aquatic Sciences*

- *The Open Fish Science Journal*

- *African Journal of Tropical Hydrobiology and Fisheries*

- *ICES Journal of Marine Science*

- *Reviews in Fish Biology and Fisheries*

- *International journal of fisheries and aquaculture*

- *Reviews in Fisheries Science*
    - *Chinese Fisheries Journal Listings*
    - *General Fisheries Journal Listings*

## Professional Societies

- World Council of Fisheries Societies

- American Fisheries Society

- The Fisheries Society of the British Isles

- The Japanese Society of Fisheries Science

- The Australian Society for Fish Biology

# Fisheries Management

Fisheries management draws on fisheries science in order to find ways to protect fishery resources so sustainable exploitation is possible. Modern fisheries management is often referred to as a governmental system of appropriate management rules based on defined objectives and a mix of management means to implement the rules, which are put in place by a system of monitoring control and surveillance. According to the Food and Agriculture Organization of the United Nations (FAO), there are "no clear and generally accepted definitions of fisheries management". However, the working definition used by the FAO and much cited elsewhere is:

The integrated process of information gathering, analysis, planning, consultation, decision-making, allocation of resources and formulation and implementation, with enforcement as necessary, of regulations or rules which govern fisheries activities in order to ensure the continued productivity of the resources and the accomplishment of other fisheries objectives.

## History

Fisheries have been explicitly managed in some places for hundreds of years. More than 80 percent of the worlds commercial exploitation of fish and shellfish are harvest from natural occurring populations in the oceans and freshwater areas. For example, the Māori people, New Zealand residents for about 700 years, had prohibitions against taking more than what could be eaten and about giving back the first fish caught as an offering to sea god Tangaroa. Starting in the 18th century attempts were made to regulate fishing in the North Norwegian fishery. This resulted in the enactment of a law in 1816 on the Lofoten fishery, which established in some measure what has come to be known as territorial use rights.

"The fishing banks were divided into areas belonging to the nearest fishing base on land and further subdivided into fields where the boats were allowed to fish. The allocation of the fishing fields was in the hands of local governing committees, usually headed by the owner of the onshore facilities which the fishermen had to rent for accommodation and for drying the fish."

Governmental resource protection-based fisheries management is a relatively new idea, first developed for North European fisheries after the first Overfishing Conference held in London in 1936. In 1957 British fisheries researchers Ray Beverton and Sidney Holt published a seminal work on North Sea commercial fisheries dynamics. In the 1960s the work became the theoretical platform for North European management schemes.

After some years away from the field of fisheries management, Beverton criticized his earlier work in a paper given at the first World Fisheries Congress in Athens in 1992. "The Dynamics of Exploited Fish Populations" expressed his concerns, including the way his and Sidney Holt's work had been misinterpreted and misused by fishery biologists and managers during the previous 30 years. Nevertheless, the institutional foundation for modern fishery management had been laid.

A report by Prince Charles' International Sustainability Unit, the New York-based Environmental Defense Fund and 50in10 published in July 2014 estimated global fisheries were adding $270 billion a year to global GDP, but by full implementation of sustainable fishing, that figure could rise by an extra amount of as much as $50 billion.

## Political Objectives

According to the FAO, fisheries management should be based explicitly on political objectives, ideally with transparent priorities. Typical political objectives when exploiting a fish resource are to:

- maximize sustainable biomass yield

- maximize sustainable economic yield

- secure and increase employment

- secure protein production and food supplies

- increase export income

Such political goals can also be a weak part of fisheries management, since the objectives can conflict with each other.

## International Objectives

Fisheries objectives need to be expressed in concrete management rules. In most countries fisheries management rules should be based on the internationally agreed, though non-binding, Code of Conduct for Responsible Fisheries, agreed at a meeting of the U.N.'s Food and Agriculture Organization FAO session in 1995. The precautionary approach it prescribes is typically implemented in concrete management rules as minimum spawning biomass, maximum fishing mortality rates, etc. In 2005 the UBC Fisheries Centre at the University of British Columbia comprehensively reviewed the performance of the world's major fishing nations against the Code.

International agreements are required in order to regulate fisheries in international waters. The desire for agreement on this and other maritime issues led to three conferences on the Law of the Sea, and ultimately to the treaty known as the United Nations Convention on the Law of the Sea (UNCLOS). Concepts such as exclusive economic zones (EEZ, extending 200 nautical miles (370 km) from a nation's coasts) allocate certain sovereign rights and responsibilities for resource management to individual countries.

Other situations need additional intergovernmental coordination. For example, in the Mediterranean Sea and other relatively narrow bodies of water, EEZ of 200 nautical miles (370 km) are irrelevant. International waters beyond 12-nautical-mile (22 km) from shore require explicit agreements.

Straddling fish stocks, which migrate through more than one EEZ also present challenges. Here sovereign responsibility must be agreed with neighbouring coastal states and fishing entities. Usually this is done through the medium of a regional organisation set up for the purpose of coordinating the management of that stock.

UNCLOS does not prescribe precisely how fisheries confined only to international waters should be managed. Several new fisheries (such as high seas bottom trawling fisheries) are not (yet) subject to international agreement across their entire range. In November 2004 the UN General Assembly issued a resolution on Fisheries that prepared for further development of international fisheries management law.

## Management Mechanisms

Many countries have set up Ministries/Government Departments, named "Ministry of Fisheries" or similar, controlling aspects of fisheries within their exclusive economic zones. Four categories of management means have been devised, regulating either input/investment, or output, and operating either directly or indirectly:

|         | Inputs          | Outputs                             |
|---------|-----------------|-------------------------------------|
| Indirect | Vessel licensing | Catching techniques                 |
| Direct  | Limited entry   | Catch quota and technical regulation |

Technical means may include:

- prohibiting devices such as bows and arrows, and spears, or firearms

- prohibiting nets

- setting minimum mesh sizes

- limiting the average potential catch of a vessel in the fleet (vessel and crew size, gear, electronic gear and other physical "inputs".

- prohibiting bait

- snagging

- limits on fish traps

- limiting the number of poles or lines per fisherman

- restricting the number of simultaneous fishing vessels

- limiting a vessel's average operational intensity per unit time at sea

- limiting average time at sea

## Catch Quotas

Systems that use *individual transferable quotas* (ITQ), also called individual fishing quota limit the total catch and allocate shares of that quota among the fishers who work that fishery. Fishers can buy/sell/trade shares as they choose.

A large scale study in 2008 provided strong evidence that ITQ's can help to prevent fishery collapse and even restore fisheries that appear to be in decline. Other studies have shown negative socio-economic consequences of ITQs, especially on small-sclale fisheries. These consequences include concentration of quota in that hands of few fishers; increased number of inactive fishers leasing their quotas to others (a phenomenon known as armchair fishermen); and detrimental effects on coastal communities.

## Precautionary Principle

The *Fishery Manager's Guidebook* issued in 2009 by the FAO of the United Nations, advises that the precautionary approach or principle should be applied when "ecosystem resilience and human impact (including reversibility) are difficult to forecast and hard to distinguish from natural changes." The precautionary principle suggests that when an action risks harm, it should not be proceeded with until it can be scientifically proven to be safe. Historically fishery managers have applied this principle the other way round; fishing activities have not been curtailed until it has been proven that they have already damaged existing ecosystems. In a paper published in 2007, Shertzer and Prager suggested that there can be significant benefits to stock biomass and fishery yield if management is stricter and more prompt.

## Fisheries Law

Fisheries law is an emerging and specialized area of law which includes the study and analysis of different fisheries management approaches, including seafood safety regulations and aquaculture regulations. Despite its importance, this area is rarely taught at law schools around the world, which leaves a vacuum of advocacy and research.

## Climate Change

In the past, changing climate has affected inland and offshore fisheries and such changes are likely to continue. From a fisheries perspective, the specific driving factors of climate change include rising water temperature, alterations in the hydrologic cycle, changes in nutrient fluxes, and relocation of spawning and nursery habitat. Further, changes in such factors would affect resources at all levels of biological organization, including the genetic, organism, population, and ecosystem levels.

## Population Dynamics

Population dynamics describes the growth and decline of a given fishery stock over time, as controlled by birth, death and migration. It is the basis for understanding changing fishery patterns and issues such as habitat destruction, predation and optimal harvesting rates. The population dynamics of fisheries has been traditionally used by fisheries scientists to determine sustainable yields.

The basic accounting relation for population dynamics is the BIDE model:

$$N_1 = N_0 + B - D + I - E$$

where $N_1$ is the number of individuals at time 1, $N_0$ is the number of individuals at time 0, $B$ is the number of individuals born, $D$ the number that died, $I$ the number that immigrated, and $E$ the number that emigrated between time 0 and time 1. While immigration and emigration can be present in wild fisheries, they are usually not measured.

Care is needed when applying population dynamics to real world fisheries. In the past, over-simplistic modelling, such as ignoring the size, age and reproductive status of the fish, focusing solely on a single species, ignoring bycatch and physical damage to the ecosystem, has accelerated the collapse of key stocks.

## Ecosystem Based Fisheries

*"We propose that rebuilding ecosystems, and not sustainability per se, should be the goal of fishery management. Sustainability is a deceptive goal because human harvesting of fish leads to a progressive simplification of ecosystems in favour of smaller, high turnover, lower trophic level fish species that are adapted to withstand disturbance and habitat degradation. "*

*— Tony Pitcher and Daniel Pauly,*

According to marine ecologist Chris Frid, the fishing industry points to pollution and global warming as the causes of unprecedentedly low fish stocks in recent years, writing, "Everybody would like to see the rebuilding of fish stocks and this can only be achieved if we understand all of the influences, human and natural, on fish dynamics." Overfishing has also had an effect. Frid adds, "Fish communities can be altered in a number of ways, for example they can decrease if particular sized individuals of a species are targeted, as this affects predator and prey dynamics. Fishing, however, is not the sole perpetrator of changes to marine life - pollution is another example [...] No one factor operates in isolation and components of the ecosystem respond differently to each individual factor."

In contrast to the traditional approach of focusing on a single species, the ecosystem-based approach is organized in terms of ecosystem services. Ecosystem-based fishery concepts have been implemented in some regions. In 2007 a group of scientists offered the following *ten commandments*

"Keep a perspective that is holistic, risk-adverse and adaptive.

- Maintain an "old growth" structure in fish populations, since big, old and fat female fish have been shown to be the best spawners, but are also susceptible to overfishing.

- Characterize and maintain the natural spatial structure of fish stocks, so that management boundaries match natural boundaries in the sea.

- Monitor and maintain seafloor habitats to make sure fish have food and shelter.

- Maintain resilient ecosystems that are able to withstand occasional shocks.

- Identify and maintain critical food-web connections, including predators and forage species.

- Adapt to ecosystem changes through time, both short-term and on longer cycles of decades or centuries, including global climate change.

- Account for evolutionary changes caused by fishing, which tends to remove large, older fish.

- Include the actions of humans and their social and economic systems in all ecological equations."

Report to Congress (2009): The State of Science to Support an Ecosystem Approach to Regional Fishery Management National Marine Fisheries Service, NOAA Technical Memorandum NMFS-F/SPO-96.

## Elderly Maternal Fish

Traditional management practices aim to reduce the number of old, slow-growing fish, leaving more room and resources for younger, faster-growing fish. Most marine fish produce huge numbers of eggs. The assumption was that younger spawners would produce plenty of viable larvae.

However, 2005 research on rockfish shows that large, elderly females are far more important than younger fish in maintaining productive fisheries. The larvae produced by these older maternal fish grow faster, survive starvation better, and are much more likely to survive than the offspring of younger fish. Failure to account for the role of older fish may help explain recent collapses of some major US West Coast fisheries. Recovery of some stocks is expected to take decades. One way to prevent such collapses is to establish marine reserves, where fishing is not allowed and fish populations age naturally.

Old fat female rockfish are the best producers

## Data Quality

According to fisheries scientist Milo Adkison, the primary limitation in fisheries management decisions is the absence of quality data. Fisheries management decisions are often based on population models, but the models need quality data to be effective. He asserts that scientists and fishery managers would be better served with simpler models and improved data.

The most reliable source for summary statistics is the FAO Fisheries Department.

## Ecopath

Ecopath, with Ecosim (EwE), is an ecosystem modelling software suite. It was initially a NOAA initiative led by Jeffrey Polovina, later primarily developed at the UBC Fisheries Centre of the University of British Columbia. In 2007, it was named as one of the ten biggest scientific breakthroughs in NOAA's 200-year history. The citation states that Ecopath "revolutionized scientists' ability worldwide to understand complex marine ecosystems". Behind this lies two decades of development work by Villy Christensen, Carl Walters, Daniel Pauly, and other fisheries scientists. As of 2010 there are 6000 registered users in 155 countries. Ecopath is widely used in fisheries management as a tool for modelling and visualising the complex relationships that exist in real world marine ecosystems.

## Human Factors

Managing fisheries is about managing people and businesses, and not about managing fish. Fish populations are managed by regulating the actions of people. If fisheries management is to be successful, then associated human factors, such as the reactions of fishermen, are of key importance, and need to be understood.

Management regulations must also consider the implications for stakeholders. Commercial fishermen rely on catches to provide for their families just as farmers rely on crops. Commercial fishing can be a traditional trade passed down from generation to generation. Most commercial fishing is based in towns built around the fishing industry; regulation changes can impact an entire town's economy. Cuts in harvest quotas can have adverse effects on the ability of fishermen to compete with the tourism industry.

## Performance

The biomass of global fish stocks has been allowed to run down. This biomass is now diminished to the point where it is no longer possible to sustainably catch the amount of fish that could be caught. According to a 2008 UN report, titled *The Sunken Billions: The Economic Justification for Fisheries Reform*, the world's fishing fleets incur a "$US 50 billion annual economic loss" through depleted stocks and poor fisheries management. The report, produced jointly by the World Bank and the UN Food and Agriculture Organization (FAO), asserts that half the world's fishing fleet could be scrapped with no change in catch.

"By improving governance of marine fisheries, society could capture a substantial part of this $50 billion annual economic loss. Through comprehensive reform, the fisheries sector could become a basis for economic growth and the creation of alternative livelihoods in many countries. At the same time, a nation's natural capital in the form of fish stocks could be greatly increased and the negative impacts of the fisheries on the marine environment reduced."

The most prominent failure of fisheries management in recent times has perhaps been the events that lead to the collapse of the northern cod fisheries. More recently, the International Consortium of Investigative Journalists produced a series of journalistic investigations called *Looting the seas*. These detail investigations into the black market for bluefin tuna, the subsidies propping up the Spanish fishing industry, and the overfishing of the Chilean jack mackerel.

## References

- Hart, Paul J B and Reynolds, John D (2002) Handbook of Fish Biology and Fisheries, Chapter 1, The human dimensions of fisheries science. Blackwell Publishing. ISBN 0-632-06482-X

- Megrey BA and Moksness E (eds) (2009) Computers in Fisheries Research second edition, Springer. ISBN 978-1-4020-8635-9. doi:10.1007/978-1-4020-8636-6_1

- Payne A, Cotter AJR, Cotter J and Potter T (2008) Advances in fisheries science: 50 years on from Beverton and Holt John Wiley and Sons. ISBN 978-1-4051-7083-3.

- Arnason, R; Kelleher, K; Willmann, R (2008). The Sunken Billions: The Economic Justification for Fisheries Reform. World Bank and FAO. ISBN 978-0-8213-7790-1.

- Beverton, R. J. H.; Holt, S. J. (1957). On the Dynamics of Exploited Fish Populations. Fishery Investigations Series II Volume XIX. Chapman and Hall (Blackburn Press, 2004). ISBN 978-1-930665-94-1.

- Caddy JF and Mahon R (1995) "Reference points for fisheries management" FAO Fisheries technical paper 347, Rome. ISBN 92-5-103733-7

- McGoodwin JR (2001) Understanding the cultures of fishing communities. A key to fisheries management and food security FAO Fisheries, Technical Paper 401. ISBN 978-92-5-104606-7.

- Morgan, Gary; Staples, Derek and Funge-Smith, Simon (2007) Fishing capacity management and illegal, unreported and unregulated fishing in Asia FAO RAP Publication. 2007/17. ISBN 978-92-5-005669-2

- Townsend, R; Shotton, Ross and Uchida, H (2008) Case studies in fisheries self-governance FAO Fisheries Technical Paper. No 504. ISBN 978-92-5-105897-8

- Walters, Carl J. and Steven J. D. Martell (2004) Fisheries ecology and management Princeton University Press. ISBN 978-0-691-11545-0.

# Fisheries Harvesting

Concerns over climate change and the propagation of biodiversity have led to the division of fisheries into sustainable fisheries and wild fisheries. The chapter strategically encompasses and incorporates the major components and key concepts of fisheries, providing a complete understanding.

## Sustainable Fishery

A conventional idea of a sustainable fishery is that it is one that is harvested at a sustainable rate, where the fish population does not decline over time because of fishing practices. Sustainability in fisheries combines theoretical disciplines, such as the population dynamics of fisheries, with practical strategies, such as avoiding overfishing through techniques such as individual fishing quotas, curtailing destructive and illegal fishing practices by lobbying for appropriate law and policy, setting up protected areas, restoring collapsed fisheries, incorporating all externalities involved in harvesting marine ecosystems into fishery economics, educating stakeholders and the wider public, and developing independent certification programs.

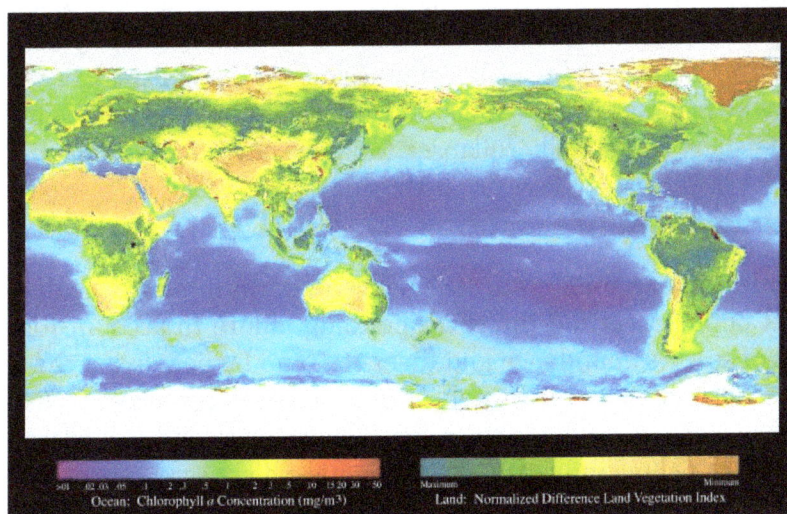

SeaWiFS map showing the levels of primary production in the world's oceans

Some primary concerns around sustainability are that heavy fishing pressures, such as overexploitation and growth or recruitment overfishing, will result in the loss of significant potential yield; that stock structure will erode to the point where it loses diversity and resilience to environmental fluctuations; that ecosystems and their economic infrastructures will cycle between collapse and recovery; with each cycle less productive than its predecessor; and that changes will occur in the trophic balance (fishing down marine food webs).

## Overview

*"Sustainable management of fisheries cannot be achieved without an acceptance that the long-term goals of fisheries management are the same as those of environmental conservation"*

*— Daniel Pauly and Dave Preikshot,*

Global wild fisheries are believed to have peaked and begun a decline, with valuable habitats, such as estuaries and coral reefs, in critical condition. Current aquaculture or farming of piscivorous fish, such as salmon, does not solve the problem because farmed piscivores are fed products from wild fish, such as forage fish. Salmon farming also has major negative impacts on wild salmon. Fish that occupy the higher trophic levels are less efficient sources of food energy.

Fishery ecosystems are an important subset of the wider marine environment. This article documents the views of fisheries scientists and marine conservationists about innovative approaches towards sustainable fisheries.

## History

*"In the end, we will conserve only what we love; we will love only what we understand; and we will understand only what we are taught"*

*— Senegalese conservationist Baba Dioum,*

In his 1883 inaugural address to the International Fisheries Exhibition in London, Thomas Huxley asserted that overfishing or "permanent exhaustion" was scientifically impossible, and stated that probably "all the great sea fisheries are inexhaustible". In reality, by 1883 marine fisheries were already collapsing. The United States Fish Commission was established 12 years earlier for the purpose of finding why fisheries in New England were declining. At the time of Huxley's address, the Atlantic halibut fishery had already collapsed (and has never recovered).

## Traditional Management of Fisheries

Traditionally, fisheries management and the science underpinning it was distorted by its "narrow focus on target populations and the corresponding failure to account for ecosystem effects leading to declines of species abundance and diversity" and by perceiving the fishing industry as "the sole legitimate user, in effect the owner, of marine living resources." Historically, stock assessment scientists usually worked in government laboratories and considered their work to be providing services to the fishing industry. These scientists dismissed conservation issues and distanced themselves from the scientists and the science that raised the issues. This happened even as commercial fish stocks deteriorated, and even though many governments were signatories to binding conservation agreements.

## Defining Sustainability

The notion of sustainable development is sometimes regarded as an unattainable, even illogical notion because development inevitably depletes and degrades the environment.

Ray Hilborn, of the University of Washington, distinguishes three ways of defining a sustainable fishery:

- *Long term constant yield* is the idea that undisturbed nature establishes a steady state that changes little over time. Properly done, fishing at up to maximum sustainable yield allows nature to adjust to a new steady state, without compromising future harvests. However, this view is naive, because constancy is not an attribute of marine ecosystems, which dooms this approach. Stock abundance fluctuates naturally, changing the potential yield over short and long term periods.

- *Preserving intergenerational equity* acknowledges natural fluctuations and regards as unsustainable only practices which damage the genetic structure destroy habitat, or deplete stock levels to the point where rebuilding requires more than a single generation. Providing rebuilding takes only one generation, overfishing may be economically foolish, but it is not unsustainable. This definition is widely accepted.

- *Maintaining a biological, social and economic system* considers the health of the human ecosystem as well as the marine ecosystem. A fishery which rotates among multiple species can deplete individual stocks and still be sustainable so long as the ecosystem retains its intrinsic integrity. Such a definition might consider as sustainable fishing practices that lead to the reduction and possible extinction of some species.

## Social Sustainability

Fisheries and aquaculture are, directly or indirectly, a source of livelihood for over 500 million people, mostly in developing countries.

Social sustainability can conflict with biodiversity. A fishery is socially sustainable if the fishery ecosystem maintains the ability to deliver products the society can use. Major species shifts within the ecosystem could be acceptable as long as the flow of such products continues. Humans have been operating such regimes for thousands of years, transforming many ecosystems, depleting or driving to extinction many species.

*"To a great extent, sustainability is like good art, it is hard to describe but we know it when we see it."*

*— Ray Hilborn,*

According to Hilborn, the "loss of some species, and indeed transformation of the ecosystem is not incompatible with sustainable harvests." For example, in recent years, barndoor skates have been caught as bycatch in the western Atlantic. Their numbers have severely declined and they will probably go extinct if these catch rates continue. Even if the barndoor skate goes extinct, changing the ecosystem, there could still be sustainable fishing of other commercial species.

## Reconciling Fisheries with Conservation

At the Fourth World Fisheries Congress in 2004, Daniel Pauly asked, "How can fisheries science and conservation biology achieve a reconciliation?", then answered his own question, "By accepting each other's essentials: that fishing should remain a viable occupation; and that aquatic ecosystems and their biodiversity are allowed to persist."

Management goals might consider the impact of salmon on bear and river ecosystems

A relatively new concept is relationship farming. This is a way of operating farms so they restore the food chain in their area. Re-establishing a healthy food chain can result in the farm automatically filtering out impurities from feed water and air, feeding its own food chain, and additionally producing high net yields for harvesting. An example is the large cattle ranch Veta La Palma in southern Spain. Relationship farming was first made popular by Joel Salatin who created a 220 hectare relationship farm featured prominently in Michael Pollan's book *The Omnivore's Dilemma* (2006) and the documentary films, Food, Inc. and Fresh. The basic concept of relationship farming is to put effort into building a healthy food chain, and then the food chain does the hard work.

## Obstacles

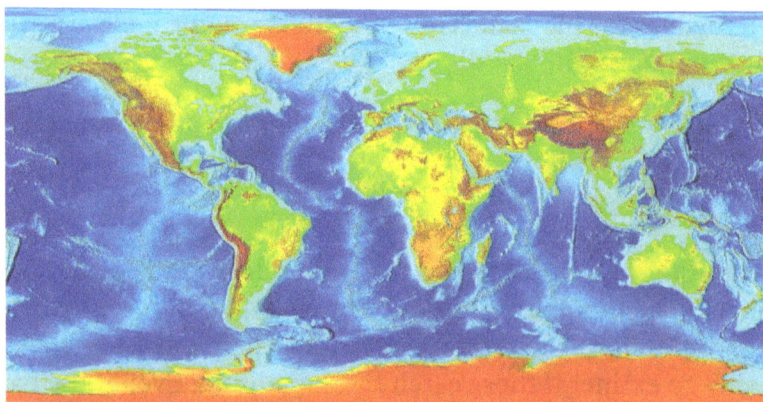

Large areas of the global continental shelf, highlighted in cyan, have had heavy
bottom trawls repeatedly dragged over them

## Overfishing

Overfishing can be sustainable. According to Hilborn, overfishing can be "a misallocation of societies' resources", but it does not necessarily threaten conservation or sustainability".

Overfishing is traditionally defined as harvesting so many fish that the yield is less than it would be if fishing were reduced. For example, Pacific salmon are usually managed by trying to determine how many spawning salmon, called the "escapement", are needed each generation to produce the maximum harvestable surplus. The optimum escapement is that needed to reach that surplus. If the escapement is half the optimum, then normal fishing looks like overfishing. But this is still sustainable fishing, which could continue indefinitely at its reduced stock numbers and yield. There is a wide range of escapement sizes that present no threat that the stock might collapse or that the stock structure might erode.

Fishing down the food web

On the other hand, overfishing can precede severe stock depletion and fishery collapse. Hilborn points out that continuing to exert fishing pressure while production decreases, stock collapses and the fishery fails, is largely "the product of institutional failure."

Coastal fishing communities in Bangladesh are vulnerable to flooding from sea-level rises.

Today over 70% of fish species are either fully exploited, overexploited, depleted, or recovering from depletion. If overfishing does not decrease, it is predicted that stocks of all species currently commercially fished for will collapse by 2048."

A Hubbert linearization (Hubbert curve) has been applied to the whaling industry, as well as charting the price of caviar, which depends on sturgeon stocks. Another example is North Sea cod. Comparing fisheries and mineral extraction tells us that human pressure on the environment is causing a wide range of resources to go through a Hubbert depletion cycle.

Island with fringing reef in the Maldives. Coral reefs are dying around the world.

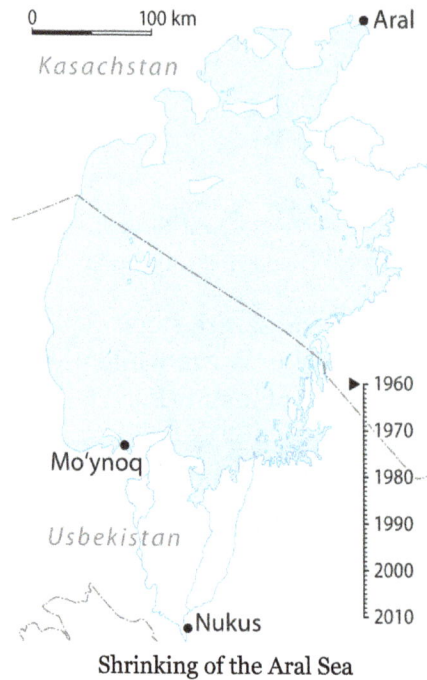

Shrinking of the Aral Sea

## Habitat Modification

Nearly all the world's continental shelves, and large areas of continental slopes, underwater ridges, and seamounts, have had heavy bottom trawls and dredges repeatedly dragged over their surfaces. For fifty years, governments and organizations, such as the Asian Development Bank, have encouraged the fishing industry to develop trawler fleets. Repeated bottom trawling and dredging literally flattens diversity in the benthic habitat, radically changing the associated communities.

## Changing the Ecosystem Balance

Since 1950, 90 percent of 25 species of big predator fish have gone.

- How we are emptying our seas *The Sunday Times*, May 10, 2009.

- Pauly, Daniel (2004) Reconciling Fisheries with Conservation: the Challenge of Managing Aquatic Ecosystems Fourth World Fisheries Congress, Vancouver, 2004.

## Climate Change

Rising ocean temperatures and ocean acidification are radically altering aquatic ecosystems. Climate change is modifying fish distribution and the productivity of marine and freshwater species. This reduces sustainable catch levels across many habitats, puts pressure on resources needed for aquaculture, on the communities that depend on fisheries, and on the oceans' ability to capture and store carbon (biological pump). Sea level rise puts coastal fishing communities at risk, while changing rainfall patterns and water use impact on inland (freshwater) fisheries and aquaculture.

## Ocean Pollution

A recent survey of global ocean health concluded that all parts of the ocean have been impacted by human development and that 41 percent has been fouled with human polluted runoff, overfishing, and other abuses. Pollution is not easy to fix, because pollution sources are so dispersed, and are built into the economic systems we depend on.

The United Nations Environment Programme (UNEP) mapped the impacts of stressors such as climate change, pollution, exotic species, and over-exploitation of resources on the oceans. The report shows at least 75 percent of the world's key fishing grounds may be affected.

## Diseases and Toxins

Large predator fish contain significant amounts of mercury, a neurotoxin which can affect fetal development, memory, mental focus, and produce tremors.

## Irrigation

Abandoned ship near Aral, Kazakhstan.

Lakes are dependent on the inflow of water from its drainage basin. In some areas, aggressive irrigation has caused this inflow to decrease significantly, causing water depletion and a shrinking of the lake. The most notable example is the Aral Sea, formerly among the four largest lakes in the world, now only a tenth of its former surface area.

## Remediation

### Fisheries Management

Fisheries management draws on fisheries science to enable sustainable exploitation. Modern fisheries management is often defined as mandatory rules based on concrete objectives and a mix of management techniques, enforced by a monitoring control and surveillance system.

- Ideas and rules: Economist Paul Romer believes sustainable growth is possible providing the right ideas (technology) are combined with the right rules, rather than simply hectoring fishers. There has been no lack of innovative ideas about how to harvest fish. He characterizes failures as primarily failures to apply appropriate rules.

- Fishing subsidies: Government subsidies influence many of the world fisheries. Operating cost subsidies allow European and Asian fishing fleets to fish in distant waters, such as West Africa. Many experts reject fishing subsidies and advocate restructuring incentives globally to help struggling fisheries recover.

- Economics: Another focus of conservationists is on curtailing detrimental human activities by improving fisheries' market structure with techniques such as salable fishing quotas, like those set up by the Northwest Atlantic Fisheries Organization, or laws such as those listed below.

- Payment for Ecosystem Services: Environmental Economist, Essam Y Mohammed, argues that by creating direct economic incentives, whereby people are able to receive payment for the services their property provides, will help to establish sustainable fisheries around the world as well as inspire conservation where it otherwise would not.

- Sustainable fisheries certification: A promising direction is the independent certification programs for sustainable fisheries conducted by organizations such as the Marine Stewardship Council and Friend of the Sea. These programs work at raising consumer awareness and insight into the nature of their seafood purchases.

- Ecosystem based fisheries.

### Ecosystem Based Fisheries

*"We propose that rebuilding ecosystems, and not sustainability per se, should be the goal of fishery management. Sustainability is a deceptive goal because human harvesting of fish leads to a progressive simplification of ecosystems in favour of smaller, high turnover, lower trophic level fish species that are adapted to withstand disturbance and habitat degradation."*

*— Tony Pitcher and Daniel Pauly,*

According to marine ecologist Chris Frid, the fishing industry points to marine pollution and global warming as the causes of recent, unprecedented declines in fish populations. Frid counters that overfishing has also altered the way the ecosystem works. "Everybody would like to see the rebuilding of fish stocks and this can only be achieved if we understand all of the influences, human and natural, on fish dynamics." He adds: "fish communities can be altered in a number of ways, for example they can decrease if particular-sized individuals of a species are targeted, as this affects predator and prey dynamics. Fishing, however, is not the sole cause of changes to marine life—pollution is another example....No one factor operates in isolation and components of the ecosystem respond differently to each individual factor."

The traditional approach to fisheries science and management has been to focus on a single species. This can be contrasted with the ecosystem-based approach. Ecosystem-based fishery concepts have been implemented in some regions. In a 2007 effort to "stimulate much needed discussion" and "clarify the essential components" of ecosystem-based fisheries science, a group of scientists offered the following ten commandments for ecosystem-based fisheries scientists

"Keep a perspective that is holistic, risk-adverse and adaptive.

- Maintain an "old growth" structure in fish populations, since big, old and fat female fish have been shown to be the best spawners, but are also susceptible to overfishing.

- Characterize and maintain the natural spatial structure of fish stocks, so that management boundaries match natural boundaries in the sea.

- Monitor and maintain seafloor habitats to make sure fish have food and shelter.

- Maintain resilient ecosystems that are able to withstand occasional shocks.

- Identify and maintain critical food-web connections, including predators and forage species.

- Adapt to ecosystem changes through time, both short-term and on longer cycles of decades or centuries, including global climate change.

- Account for evolutionary changes caused by fishing, which tends to remove large, older fish.

- Include the actions of humans and their social and economic systems in all ecological equations." Marine protected areas

Strategies and techniques for marine conservation tend to combine theoretical disciplines, such as population biology, with practical conservation strategies, such as setting up protected areas, as with Marine Protected Areas (MPAs) or Voluntary Marine Conservation Areas. Each nation defines MPAs independently, but they commonly involve increased protection for the area from fishing and other threats.

Marine life is not evenly distributed in the oceans. Most of the really valuable ecosystems are in relatively shallow coastal waters, above or near the continental shelf, where the sunlit waters are often nutrient rich from land runoff or upwellings at the continental edge, allowing photosynthesis, which energizes the lowest trophic levels. In the 1970s, for reasons more to do with oil drilling than with fishing, the U.S. extended its jurisdiction, then 12 miles from the coast, to 200 miles. This made huge shelf areas part of its territory. Other nations followed, extending national control to what became known as the exclusive economic zone (EEZ). This move has had many implications for fisheries conservation, since it means that most of the most productive maritime ecosystems are now under national jurisdictions, opening possibilities for protecting these ecosystems by passing appropriate laws.

Daniel Pauly characterises marine protected areas as "a conservation tool of revolutionary importance that is being incorporated into the fisheries mainstream." The Pew Charitable Trusts have funded various initiatives aimed at encouraging the development of MPAs and other ocean conservation measures.

## Fish Farming

There exists concerns that farmed fish cannot produce necessary yields efficiently. For example, farmed salmon eat three pounds of wild fish to produce one pound of salmon.

## Laws and Treaties

International laws and treaties related to marine conservation include the 1966 Convention on Fishing and Conservation of Living Resources of the High Seas. United States laws related to marine conservation include the 1972 Marine Mammal Protection Act, as well as the 1972 Marine Protection, Research and Sanctuaries Act which established the National Marine Sanctuaries program. Magnuson-Stevens Fishery Conservation and Management Act.

## Awareness Campaigns

Various organizations promote sustainable fishing strategies, educate the public and stakeholders, and lobby for conservation law and policy. The list includes the Marine Conservation Biology Institute and Blue Frontier Campaign in the U.S., The U.K.'s Frontier (the Society for Environmental Exploration) and Marine Conservation Society, Australian Marine Conservation Society, International Council for the Exploration of the Sea (ICES), Langkawi Declaration, Oceana, PROFISH, and the Sea Around Us Project, International Collective in Support of Fishworkers, World Forum of Fish Harvesters and Fish Workers, Frozen at Sea Fillets Association and CEDO.

Introducing the results of long term monitoring to a local fishermen in Kihnu, Estonia.

The United Nations Millennium Development Goals include, as goal #7: target 2, the intention to "reduce biodiversity loss, achieving, by 2010, a significant reduction in the rate of loss", including improving fisheries management to reduce depletion of fish stocks.

Some organizations certify fishing industry players for sustainable or good practices, such as the Marine Stewardship Council and Friend of the Sea.

Other organizations offer advice to members of the public who eat with an eye to sustainability. According to the marine conservation biologist Callum Roberts, four criteria apply when choosing seafood:

- Is the species in trouble in the wild where the animals were caught?

- Does fishing for the species damage ocean habitats?

- Is there a large amount of bycatch taken with the target species?

- Does the fishery have a problem with discards—generally, undersized animals caught and thrown away because their market value is low?

The following organizations have download links for wallet-sized cards, listing good and bad choices:

- Monterey Bay Aquarium Seafood Watch, USA

- Blue Ocean Institute, USA

- Marine Conservation Society, UK

- Australian Marine Conservation Society

- The Southern African Sustainable Seafood Initiative

## Data Issues

### Data Quality

One of the major impediments to the rational control of marine resources is inadequate data. According to fisheries scientist Milo Adkison (2007), the primary limitation in fisheries management decisions is poor data. Fisheries management decisions are often based on population models, but the models need quality data to be accurate. Scientists and fishery managers would be better served with simpler models and improved data.

### Unreported Fishing

Estimates of illegal catch losses range between $10 billion and $23 billion annually, representing between 11 and 26 million tonnes.

- Incidental catch

### Shifting Baselines

Shifting baselines is a term which describes the way significant changes to a system are measured against previous baselines, which themselves may represent significant changes from the original state of the system. The term was first used by the fisheries scientist Daniel Pauly in his paper "Anecdotes and the shifting baseline syndrome of fisheries". Pauly developed the term in reference to fisheries management where fisheries scientists sometimes fail to identify the correct "baseline" population size (e.g. how abundant a fish species population was *before* human exploitation) and thus work with a shifted baseline. He describes the way that radically depleted fisheries were evaluated by experts who used the state of the fishery at the start of their careers as the baseline, rather than the fishery in its untouched state. Areas that swarmed with a particular species hundreds of years ago, may have experienced long term decline, but it is the level of decades previously that is considered the appropriate reference point for current populations. In this way large declines in ecosystems or species over long periods of time were, and are, masked. There is a loss of perception of change that occurs when each generation redefines what is "natural".

## Looting the Seas

*Looting the seas* is the name given by the International Consortium of Investigative Journalists to a series of journalistic investigations into areas directly affecting the sustainability of fisheries. So far they have investigated three areas involving fraud, negligence and overfishing:

- The black market in bluefin tuna

- Subsidies propping up the Spanish fishing industry

- Overfishing of the southern jack mackerel

# Wild Fisheries

A fishery is an area with an associated fish or aquatic population which is harvested for its commercial value. Fisheries can be marine (saltwater) or freshwater. They can also be wild or farmed.

Crab boat from the North Frisian Islands working in the North Sea

Wild fisheries are sometimes called capture fisheries. The aquatic life they support is not controlled in any meaningful way and needs to be "captured" or fished. Wild fisheries exist primarily in the oceans, and particularly around coasts and continental shelves. They also exist in lakes and rivers. Issues with wild fisheries are overfishing and pollution. Significant wild fisheries have collapsed or are in danger of collapsing, due to overfishing and pollution. Overall, production from the world's wild fisheries has levelled out, and may be starting to decline.

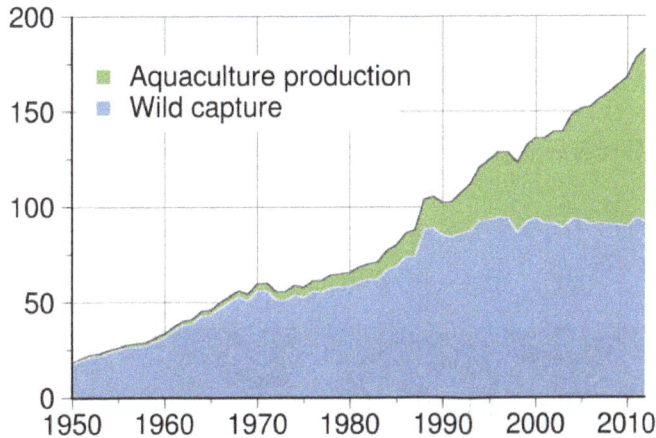

Global harvest of aquatic organisms in million tonnes, 1950–2010, as reported by the FAO

As a contrast to wild fisheries, farmed fisheries can operate in sheltered coastal waters, in rivers, lakes and ponds, or in enclosed bodies of water such as tanks. Farmed fisheries are technological in nature, and revolve around developments in aquaculture. Farmed fisheries are expanding, and Chinese aquaculture in particular is making many advances. Nevertheless, the majority of fish consumed by humans continues to be sourced from wild fisheries. As of the early 21st century, fish is humanity's only significant wild food source.

## Marine and Inland Production

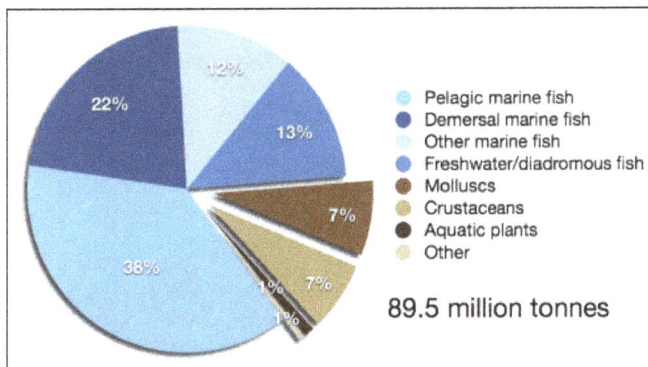

Global wild fish capture in million tonnes, 2010, as reported by the FAO

According to the Food and Agriculture Organization (FAO), the world harvest by commercial fisheries in 2010 consisted of 88.6 million tonnes of aquatic animals captured in wild fisheries, plus another 0.9 million tons of aquatic plants (seaweed etc.). This can be contrasted with 59.9 million tonnes produced in fish farms, plus another 19.0 million tons of aquatic plants harvested in aquaculture.

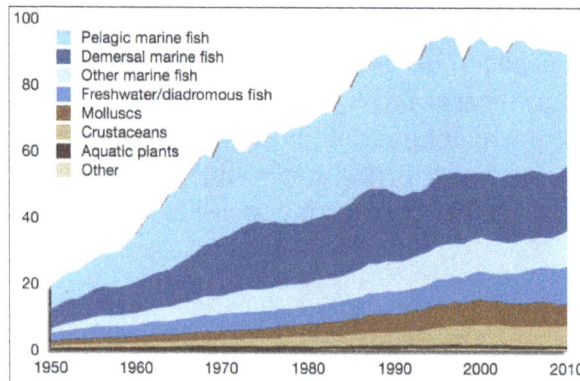

Global wild fish capture in million tonnes, 1950–2010, as reported by the FAO

# Marine Fisheries

# Topography

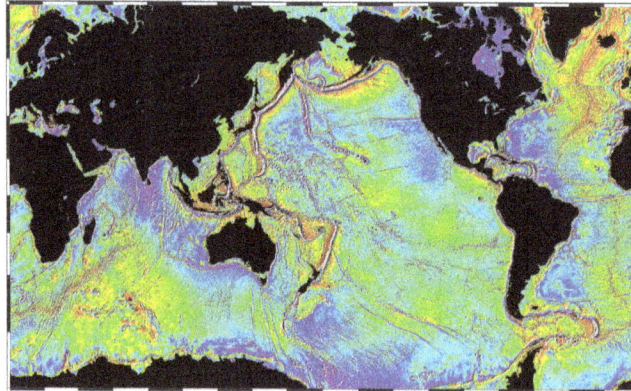

Map of underwater topography. (1995, NOAA)

The productivity of marine fisheries is largely determined by marine topography, including its interaction with ocean currents and the diminishment of sunlight with depth.

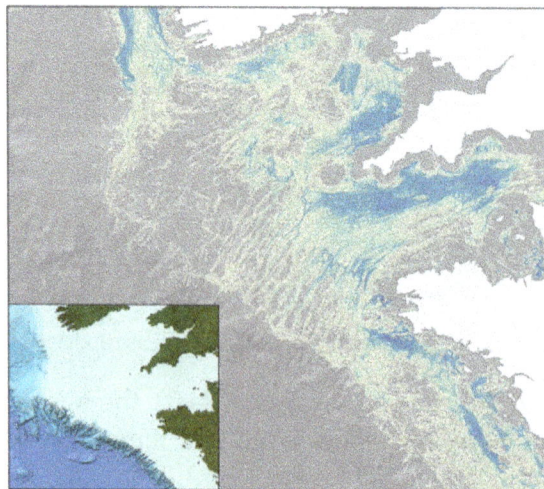

Fishing activities extracted from Automatic Identification Data of EU trawlers over the continental shelf, highlighting the correlation with the bathymetry over the area (bottom-left, from the GEBCO world map 2014).

Marine topography is defined by various coastal and oceanic landforms, ranging from coastal estuaries and shorelines; to continental shelves and coral reefs; to underwater and deep sea features such as ocean rises and seamounts.

## Ocean Currents

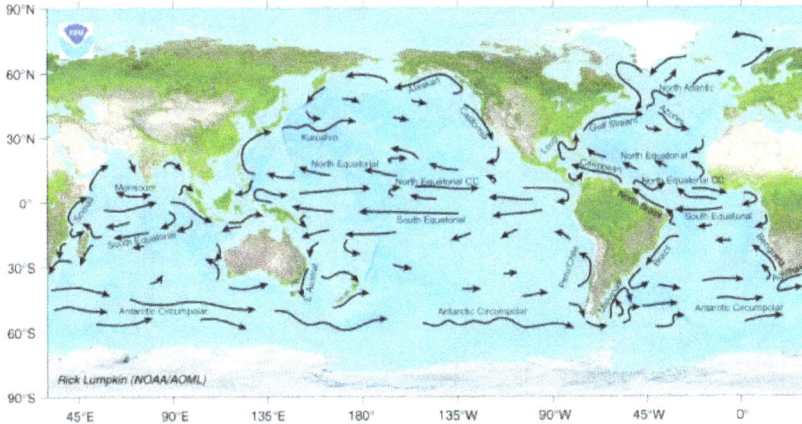

Major ocean surface currents. NOAA map.

An ocean current is continuous, directed movement of ocean water. Ocean currents are rivers of relatively warm or cold water within the ocean. The currents are generated from the forces acting upon the water like the planet rotation, the wind, the temperature and salinity (hence isopycnal) differences and the gravitation of the moon. The depth contours, the shoreline and other currents influence the current's direction and strength.

## More on Currents

Ocean currents can flow for thousands of kilometers. Surface ocean currents are generally wind driven and develop their typical clockwise spirals in the northern hemisphere and counter-clockwise rotation in the southern hemisphere because of the imposed wind stresses. In wind driven currents, the Ekman spiral effect results in the currents flowing at an angle to the driving winds. The areas of surface ocean currents move somewhat with the seasons; this is most notable in equatorial currents.

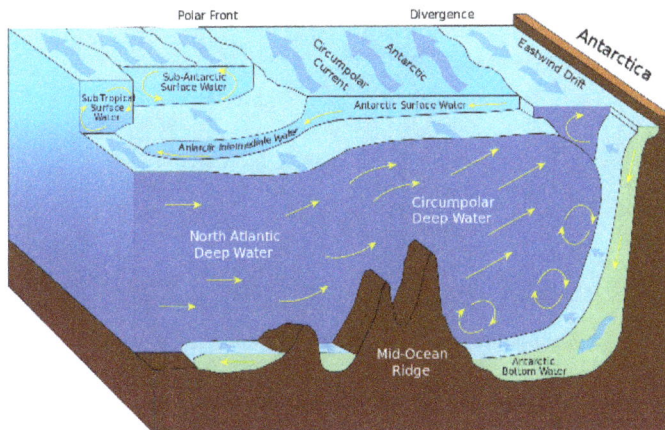

Example of different ocean currents in the Southern Ocean

Deep ocean currents are driven by density and temperature gradients. Thermohaline circulation, also known as the ocean's conveyor belt, refers to the deep ocean density-driven ocean basin currents. These currents, which flow under the surface of the ocean and are thus hidden from immediate detection, are called submarine rivers. Upwelling and downwelling areas in the oceans are areas where significant vertical movement of ocean water is observed.

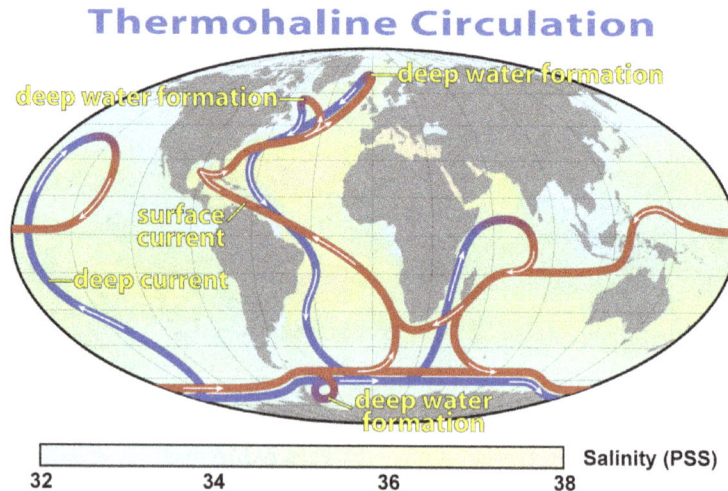

A summary of the path of the Thermohaline Circulation. Blue paths represent deep-water currents, while red paths represent surface currents

Surface currents make up about 10% of all the water in the ocean. Surface currents are generally restricted to the upper 400 meters of the ocean. The movement of deep water in the ocean basins is by density driven forces and gravity. The density difference is a function of different temperatures and salinity. Deep waters sink into the deep ocean basins at high latitudes where the temperatures are cold enough to cause the density to increase. The main causes of currents are: solar heating, winds and gravity.

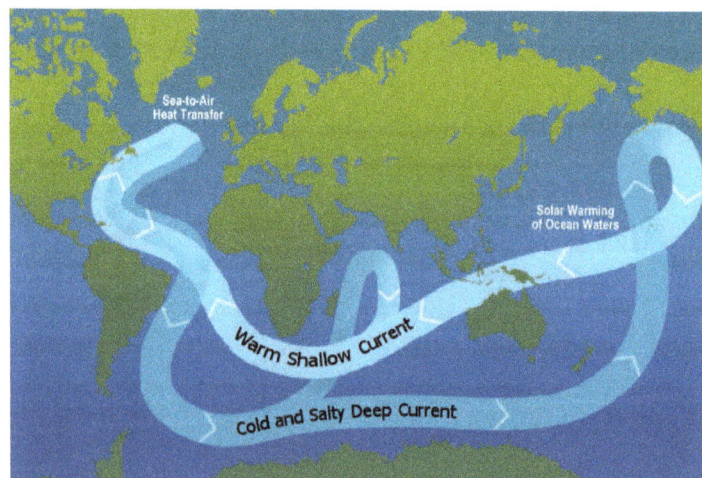

A schematic of modern thermohaline circulation

Ocean currents are also very important in the dispersal of many life forms. A dramatic example is the life-cycle of the eel. Currents also determine the disposition of marine debris. Gyres and upwelling

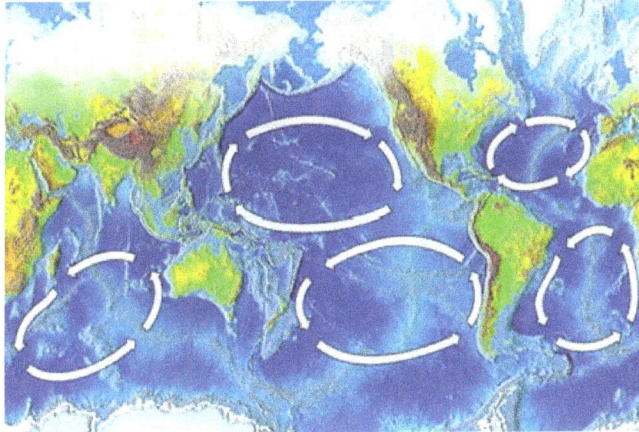

Map of Ocean Gyres

Oceanic gyres are large-scale ocean currents caused by the Coriolis effect. Wind-driven surface currents interact with these gyres and the underwater topography, such as seamounts and the edge of continental shelves, to produce downwellings and upwellings. These can transport nutrients and provide feeding grounds for plankton eating forage fish. This in turn draws larger fish that prey on the forage fish, and can result in productive fishing grounds. Most upwellings are coastal, and many of them support some of the most productive fisheries in the world, such as small pelagics (sardines, anchovies, etc.). Regions of upwelling include coastal Peru, Chile, Arabian Sea, western South Africa, eastern New Zealand and the California coast.

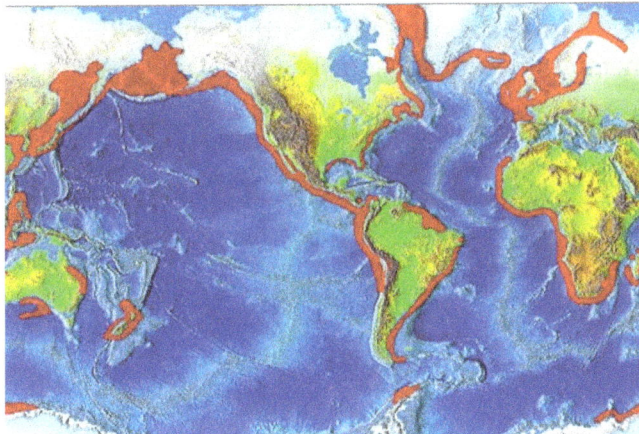

Map of regions of upwelling

## Biomass

In the ocean, the food chain typically follows the course:

- Phytoplankton→ zooplankton → predatory zooplankton → filter feeders → predatory fish

Phytoplankton is usually the primary producer (the first level in the food chain or the first trophic level). Phytoplankton converts inorganic carbon into protoplasm. Phytoplankton is consumed by microscopic animals called zooplankton. These are the second level in the food chain, and include krill, the larva of fish, squid, lobsters and crabs—as well as the small crustaceans called copepods, and many other types. Zooplankton is consumed both by other, larger predatory zooplankters and by fish (the third level in the food chain). Fish that eat zooplankton could constitute the fourth trophic level, while seals consuming the fish are the fifth. Alternatively, for example, whales may consume zooplankton directly - leading to an environment with one less trophic level.

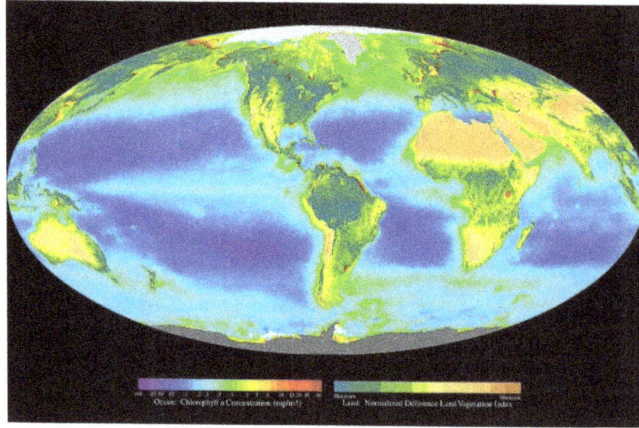

Estimate of biomass produced by photosynthesis from September 1997 to August 2000. This is a rough indicator of the primary production potential in the oceans. Provided by the SeaWiFS Project, NASA/Goddard Space Flight Center and ORBIMAGE.

## Habitats

Aquatic habitats have been classified into marine and freshwater ecoregions by the Worldwide Fund for Nature (WWF). An ecoregion is defined as a "relatively large unit of land or water containing a characteristic set of natural communities that share a large majority of their species, dynamics, and environmental conditions (Dinerstein et al. 1995, TNC 1997).

## Coastal Waters

Estuary of Klamath River

- Estuaries are semi-enclosed coastal bodies of water with one or more rivers or streams flowing into them, and with a free connection to the open sea. Estuaries are often associated with high rates of biological productivity. They are small, in demand, impacted by events far upstream or out at sea, and concentrate materials such as pollutants and sediments.

- Lagoons are bodies of comparatively shallow salt or brackish water separated from the deeper sea by a shallow or exposed sandbank, coral reef, or similar feature. *Lagoon* refers to both coastal lagoons formed by the build-up of sandbanks or reefs along shallow coastal waters, and the lagoons in atolls, formed by the growth of coral reefs on slowly sinking central islands. Lagoons that are fed by freshwater streams are estuaries.

- The intertidal zone (foreshore) is the area that is exposed to the air at low tide and submerged at high tide, for example, the area between tide marks. This area can include many different types of habitats, including steep rocky cliffs, sandy beaches or vast mudflats. The area can be a narrow strip, as in Pacific islands that have only a narrow tidal range, or can include many meters of shoreline where shallow beach slope interacts with high tidal excursion.

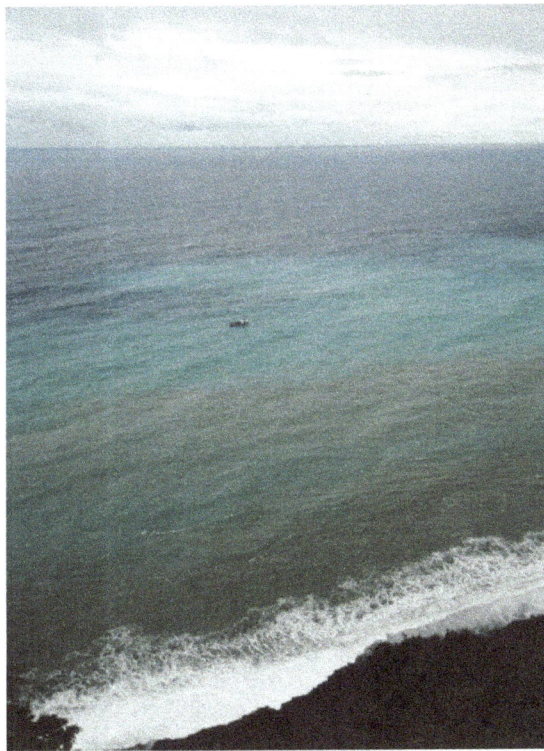

Fixed-net fishing on the littoral zone along the Suhua Highway on the East coast of Taiwan

- The littoral zone is the part of the ocean closest to the shore. The word *littoral* comes from the Latin *litoralis*, which means *seashore*. The littoral zone extends from the high-water mark to near shore areas that are permanently submerged, and includes the intertidal zone. Definitions vary. Encyclopædia Britannica defines the littoral zone in a thoroughly vague way as the "marine ecological realm that experiences the effects of tidal and longshore currents and breaking waves to a depth of 5 to 10 metres (16 to 33 feet) below the low-tide level, depending on the intensity of storm waves". The US Navy defines it as extending "from the shoreline to 600 feet (183 meters) out into the water"

- The sublittoral zone is the part of the ocean extending from the seaward edge of the littoral zone to the edge of the continental shelf. It is sometimes called the neritic zone. Websters defines the neritic zone as the region of shallow water adjoining the seacoast. The word *neritic* perhaps comes from the new Latin *nerita*, which refers to a genus of marine snails, 1891. The sublittoral zone is relatively shallow, extending to about 200 meters (100 fathoms), and generally has well-oxygenated water, low water pressure, and relatively stable temperature and salinity levels. These, combined with presence of light and the resulting photosynthetic life, such as phytoplankton and floating sargassum, make the sublittoral zone the location of the majority of sea life.

- Voigt, Brian (1998) *Glossary of Coastal Terminology* Washington State Department of Ecology, publication 98-105

- Pawson, M G; Pickett, G D and Walker, P (2002) *The coastal fisheries of England and Wales, Part IV: A review of their status 1999–2001* Science Series, Technical Report 116.

## Continental Shelves

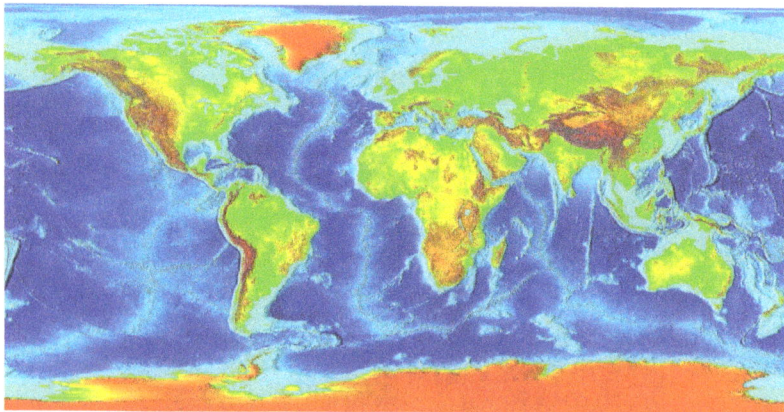

The global continental shelf, highlighted in cyan

Continental shelves are the extended perimeters of each continent and associated coastal plain, which is covered during interglacial periods such as the current epoch by relatively shallow seas (known as shelf seas) and gulfs.

The shelf usually ends at a point of decreasing slope (called the shelf break). The sea floor below the break is the continental slope. Below the slope is the continental rise, which finally merges into the deep ocean floor, the abyssal plain. The continental shelf and the slope are part of the continental margin.

Continental shelves are shallow (averaging 140 metres or 460 feet), and the sunlight available means they can teem with life. The shallowest parts of the continental shelf are called fishing banks. There the sunlight penetrates to the seafloor and the plankton, on which fish feed, thrive.

## Continental Shelves : Details

The character of the shelf changes dramatically at the shelf break, where the continental slope begins. With a few exceptions, the shelf break is located at a remarkably uniform depth of roughly 140 m (460 ft); this is likely a hallmark of past ice ages, when sea level was lower than it is now.

The width of the continental shelf varies considerably – it is not uncommon for an area to have virtually no shelf at all, particularly where the forward edge of an advancing oceanic plate dives beneath continental crust in an offshore subduction zone such as off the coast of Chile or the west coast of Sumatra. The largest shelf – the Siberian Shelf in the Arctic Ocean – stretches to 1500 kilometers (930 miles) in width. The South China Sea lies over another extensive area of continental shelf, the Sunda Shelf, which joins Borneo, Sumatra, and Java to the Asian mainland. Other familiar bodies of water that overlie continental shelves are the North Sea and the Persian Gulf. The average width of continental shelves is about 80 km (50 mi). The depth of the shelf also varies, but is generally limited to water shallower than 150 m (490 ft).

Combined with the sunlight available in shallow waters, the continental shelves teem with life compared to the biotic desert of the oceans' abyssal plain. The pelagic (water column) environment of the continental shelf constitutes the neritic zone, and the benthic (sea floor) province of the shelf is the sublittoral zone. Coral reefs

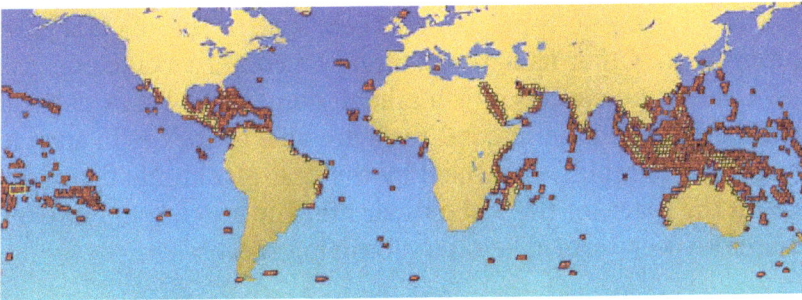

Locations of coral reefs.

Coral reefs are aragonite structures produced by living organisms, found in shallow, tropical marine waters with little to no nutrients in the water. High nutrient levels such as those found in run-off from agricultural areas can harm the reef by encouraging the growth of algae. Although corals are found both in temperate and tropical waters, reefs are formed only in a zone extending at most from 30°N to 30°S of the equator.

## Coral Reefs : Details

Coral reefs are estimated to cover 284,300 square kilometres, with the Indo-Pacific region (including the Red Sea, Indian Ocean, Southeast Asia and the Pacific) accounting for 91.9% of the total. Southeast Asia accounts for 32.3% of that figure, while the Pacific including Australia accounts for 40.8%. Atlantic and Caribbean coral reefs only account for 7.6% of the world total.

Coral reefs are either restricted or absent from the west coast of the Americas, as well as the west coast of Africa. This is due primarily to upwelling and strong cold coastal currents that reduce water temperatures in these areas. Corals are also restricted from off the coastline of South Asia from Pakistan to Bangladesh. They are also restricted along the coast around north-eastern South America and Bangladesh due to the release of vast quantities of freshwater from the Amazon and Ganges Rivers respectively.

## Famous Coral Reefs and Reef Areas of the World Include:

- The Great Barrier Reef - largest coral reef system in the world, Queensland, Australia;

- The Belize Barrier Reef - second largest in the world, stretching from southern Quintana Roo, Mexico and all along the coast of Belize down to the Bay Islands of Honduras.

- The Red Sea Coral Reef - located off the coast of Egypt and Saudi Arabia.

- Pulley Ridge - deepest photosynthetic coral reef, Florida

- Many of the numerous reefs found scattered over the Maldives

- The New Caledonia Barrier Reef - second longest double barrier reef in the world, with a length of about 1,500 km (932 mi).

Coral reefs support an extraordinary biodiversity; although they are located in nutrient-poor tropical waters. The process of nutrient cycling between corals, zooxanthellae, and other reef organisms provides an explanation for why coral reefs flourish in these waters: recycling ensures that fewer nutrients are needed overall to support the community.

Coral reefs are home to a variety of tropical or reef fish, such as the colorful parrotfish, angelfish, damselfish, and butterflyfish. Other fish groups found on coral reefs include groupers, snappers, grunts and wrasses. Over 4,000 species of fish inhabit coral reefs. It has been suggested that the high number of fish species that inhabit coral reefs are able to coexist in such high numbers because any free living space is rapidly inhabited by the first planktonic fish larvae that occupy it. These fish then inhabit the space for the rest of their life. The species that inhabit the free space is random and has therefore been termed 'a lottery for living space'.

Reefs are also home to a large variety of other organisms, including sponges, Cnidarians (which includes some types of corals and jellyfish), worms, crustaceans (including shrimp, spiny lobsters and crabs), molluscs (including cephalopods), echinoderms (including starfish, sea urchins and sea cucumbers), sea squirts, sea turtles and sea snakes.

*Bioerosion* (coral damage) such as this may be caused by coral bleaching.

Human activity may represent the greatest threat to coral reefs living in Earth's oceans. In particular, pollution and over-fishing are the most serious threats to these ecosystems. Physical destruction of reefs due to boat and shipping traffic is also a problem. The live food fish trade has been implicated as a driver of decline due to the use of cyanide and disaster for peoples living in the tropics. Hughes, et al., (2003), writes that "with increased human population and improved storage and transport systems, the scale of human impacts on reefs has grown exponentially. For example, markets for fishes and other natural resources have become global, supplying demand for reef resources far removed from their tropical sources."

Currently researchers are working to determine the degree various factors impact the reef systems. The list of factors is long but includes the oceans acting as a carbon dioxide sink, changes in Earth's atmosphere, ultraviolet light, ocean acidification, biological virus, impacts of dust storms carrying agents to far flung reef systems, various pollutants, impacts of algal blooms and others. Reefs are threatened well beyond coastal areas and so the problem is broader than factors from land development and pollution though those are too causing considerable damage.

Southeast Asian coral reefs are at risk from damaging fishing practices (such as cyanide and blast fishing), overfishing, sedimentation, pollution and bleaching. A variety of activities, including education, regulation, and the establishment of marine protected areas are under way to protect these reefs. Indonesia, for example has nearly 33,000 square miles (85,000 km²) of coral reefs. Its waters are home to a third of the world's total corals and a quarter of its fish species. Indonesia's coral reefs are located in the heart of the Coral Triangle and have been victim to destructive fishing, unregulated tourism, and bleaching due to climatic changes. Data from 414 reef monitoring stations throughout Indonesia in 2000 found that only 6% of Indonesia's coral reefs are in excellent condition, while 24% are in good condition, and approximately 70% are in poor to fair condition (2003 The Johns Hopkins University).

General estimates show approximately 10% of the coral reefs around the world are already dead. Problems range from environmental effects of fishing techniques, described above, to ocean acidification. Coral bleaching is another manifestation of the problem and is showing up in reefs across the planet.

NInhabitants of Ahus Island, Manus Province, Papua New Guinea, have followed a generations-old practice of restricting fishing in six areas of their reef lagoon. While line fishing is permitted, net and spear fishing are restricted based on cultural traditions. The result is that both the biomass and individual fish sizes are significantly larger in these areas than in places where fishing is completely unrestricted. It is estimated that about 60% of the world's reefs are at risk due to destructive, human-related activities. The threat to the health of reefs is particularly strong in Southeast Asia, where an enormous 80% of reefs are considered endangered.

Organisations as Coral Cay, Counterpart and the Foundation of the peoples of the South Pacific are currently undertaking coral reef/atoll restoration projects. They are doing so using simple methods of plant propagation. Other organisations as Practical Action have released informational documents on how to set up coral reef restoration to the public. Open sea

In the deep ocean, much of the ocean floor is a flat, featureless underwater desert called the abyssal plain. Many pelagic fish migrate across these plains in search of spawning or different feeding grounds. Smaller migratory fish are followed by larger predator fish and can provide rich, if temporary, fishing grounds.

- Strait

## Seamounts

A seamount is an underwater mountain, rising from the seafloor that does not reach to the water's surface (sea level), and thus is not an island. They are defined by oceanographers as independent features that rise to at least 1,000 meters above the seafloor. Seamounts are common in the Pacific Ocean. Recent studies suggest there may be 30,000 seamounts in the Pacific, about 1,000 in the Atlantic Ocean and an unknown number in the Indian Ocean.

The locations of the world's major seamounts

## Seamounts : Details

Seamounts often project upwards into shallower zones more hospitable to sea life, providing habitats for marine species that are not found on or around the surrounding deeper ocean bottom. In addition to simply providing physical presence in this zone, the seamount itself may deflect deep currents and create upwelling. This process can bring nutrients into the photosynthetic zone, producing an area of activity in an otherwise desert-like open ocean. Seamounts may thus be vital stopping points for some migratory animals such as whales. Some recent research indicates whales may use such features as navigational aids throughout their migration.

Due to the larger populations of fish in these areas overexpoitation by the fishing industry has caused some seamount fauna populations to decrease considerably.

The primary productivity of the epipelagic waters above the submerged peak can often be enhanced by the hydrographic conditions of the seamount. This increases the densities of the zooplankton and leads to the high concentrations of fish in these areas. Another theory for this is that the fish are sustained on the diurnal migration of zooplankton being interrupted by the presence of the seamount, and causing the zooplankton to stay in the area. It is also possible that the high densities of fishes has more to do with the fish life histories and interaction with the benthic fauna of the seamount.

The benthic fauna of the seamounts is dominated by suspension feeders, including sponges and true corals. For some seamounts that peaks at 200–300 metres below the surface benthic macroalgae is common. The sedimentary infauna is dominated by polychaete worms.

For a long time it has been surmised that many pelagic animals visit seamounts to gather food, but proof this of this aggregating effect has been lacking. The first demonstration of this conjecture has recently been published

During the 1960s, Russia, Australia and New Zealand started to look for new stocks of fish and began to trawl the seamounts. The majority of the invertebrates brought up are corals, and are mainly used for the jewelry trade. The two major fish species were the orange roughy (*Hoplostethus atlanticus*) and pelagic armourhead (*Pseudopentaceros wheeleri*), which were quickly overexploited due to lack of knowledge of the longevity of the fish, late maturity, low fecundity, small geographic range and recruitment to the fishery. As well as the fishes being overexploited the benthic communities were destroyed by the trawling gear.

- CenSeam, Census of Marine Life project CenSeam: a global census of marine life on seamounts

## Freshwater Fisheries

### Lakes

Worldwide, freshwater lakes have an area of 1.5 million square kilometres. Saline inland seas add another 1.0 million square kilometres. There are 28 freshwater lakes with an area greater than 5,000 square kilometres, totalling 1.18 million square kilometres or 79 percent of the total.

### Rivers

### Pollution

Pollution is the introduction of contaminants into an environment. Wild fisheries flourish in oceans, lakes, and rivers, and the introduction of contaminants is an issue of concern, especially as regards plastics, pesticides, heavy metals, and other industrial and agricultural pollutants which do not disintegrate rapidly in the environment. Land run-off and industrial, agricultural, and domestic waste enter rivers and are discharged into the sea. Pollution from ships is also a problem.

### Plastic Waste

Marine debris is human-created waste that ends up floating in the sea. Oceanic debris tends to accumulate at the centre of gyres and coastlines, frequently washing aground where it is known as beach litter. Eighty percent of all known marine debris is plastic - a component that has been rapidly accumulating since the end of World War II. Plastics accumulate because they don't biodegrade as many other substances do; while they will photodegrade on exposure to the sun, they do so only under dry conditions, as water inhibits this process.

Discarded plastic bags, six pack rings and other forms of plastic waste which finish up in the ocean present dangers to wildlife and fisheries. Aquatic life can be threatened through entanglement, suffocation, and ingestion.

Nurdles, also known as mermaids' tears, are plastic pellets typically under five millimetres in diameter, and are a major contributor to marine debris. They are used as a raw material in plastics manufacturing, and are thought to enter the natural environment after accidental spillages. Nurdles are also created through the physical weathering of larger plastic debris. They strongly resemble fish eggs, only instead of finding a nutritious meal, any marine wildlife that ingests them will likely starve, be poisoned and die.

Many animals that live on or in the sea consume flotsam by mistake, as it often looks similar to their natural prey. Plastic debris, when bulky or tangled, is difficult to pass, and may become permanently lodged in the digestive tracts of these animals, blocking the passage of food and causing death through starvation or infection. Tiny floating particles also resemble zooplankton, which can lead filter feeders to consume them and cause them to enter the ocean food chain. In samples taken from the North Pacific Gyre in 1999 by the Algalita Marine Research Foundation, the mass of plastic exceeded that of zooplankton by a factor of six. More recently, reports have surfaced that there may now be 30 times more plastic than plankton, the most abundant form of life in the ocean.

Toxic additives used in the manufacture of plastic materials can leech out into their surroundings when exposed to water. Waterborne hydrophobic pollutants collect and magnify on the surface of plastic debris, thus making plastic far more deadly in the ocean than it would be on land. Hydrophobic contaminants are also known to bioaccumulate in fatty tissues, biomagnifying up the food chain and putting great pressure on apex predators. Some plastic additives are known to disrupt the endocrine system when consumed, others can suppress the immune system or decrease reproductive rates.

## Toxins

Septic river.

Apart from plastics, there are particular problems with other toxins which do not disintegrate rapidly in the marine environment. Heavy metals are metallic chemical elements that have a relatively high density and are toxic or poisonous at low concentrations. Examples are mercury, lead, nickel, arsenic and cadmium. Other persistent toxins are PCBs, DDT, pesticides, furans, dioxins and phenols.

Such toxins can accumulate in the tissues of many species of aquatic life in a process called bioaccumulation. They are also known to accumulate in benthic environments, such as estuaries and bay muds: a geological record of human activities of the last century.

Some specific examples are

- Chinese and Russian industrial pollution such as phenols and heavy metals in the Amur River have devastated fish stocks and damaged its estuary soil.

- Wabamun Lake in Alberta, Canada, once the best whitefish lake in the area, now has unacceptable levels of heavy metals in its sediment and fish.

- Acute and chronic pollution events have been shown to impact southern California kelp forests, though the intensity of the impact seems to depend on both the nature of the contaminants and duration of exposure.

- Due to their high position in the food chain and the subsequent accumulation of heavy metals from their diet, mercury levels can be high in larger species such as bluefin and albacore. As a result, in March 2004 the United States FDA issued guidelines recommending that pregnant women, nursing mothers and children limit their intake of tuna and other types of predatory fish.

- Some shellfish and crabs can survive polluted environments, accumulating heavy metals or toxins in their tissues. For example, mitten crabs have a remarkable ability to survive in highly modified aquatic habitats, including polluted waters. The farming and harvesting of such species needs careful management if they are to be used as a food.

- Mining has a poor environmental track record. For example, according to the United States Environmental Protection Agency, mining has contaminated portions of the headwaters of over 40% of watersheds in the western continental US. Much of this pollution finishes up in the sea.

- Heavy metals enter the environment through oil spills - such as the Prestige oil spill on the Galician coast - or from other natural or anthropogenic sources.

Polluted lagoon.

## Eutrophication

Eutrophication is an increase in chemical nutrients, typically compounds containing nitrogen or phosphorus, in an ecosystem. It can result in an increase in the ecosystem's primary productivity (excessive plant growth and decay), and further effects including lack of oxygen and severe reductions in water quality, fish, and other animal populations.

The biggest culprit are rivers that empty into the ocean, and with it the many chemicals used as fertilizers in agriculture as well as waste from livestock and humans. An excess of oxygen depleting chemicals in the water can lead to hypoxia and the creation of a dead zone.

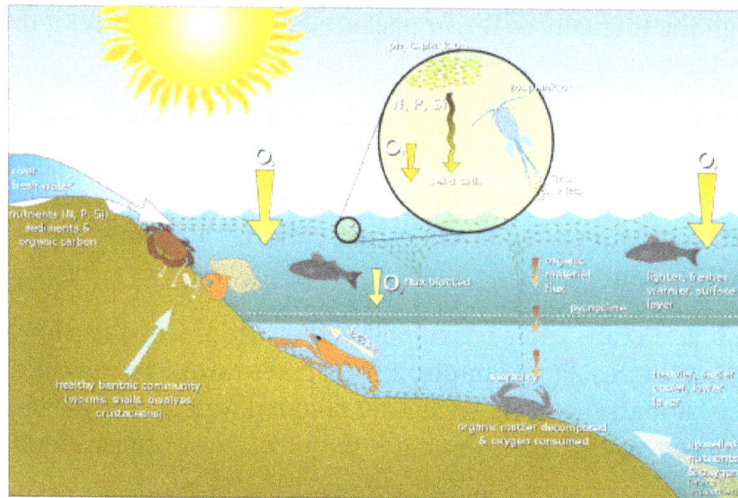

Effect of eutrophication on marine benthic life

Surveys have shown that 54% of lakes in Asia are eutrophic; in Europe, 53%; in North America, 48%; in South America, 41%; and in Africa, 28%. Estuaries also tend to be naturally eutrophic because land-derived nutrients are concentrated where run-off enters the marine environment in a confined channel. The World Resources Institute has identified 375 hypoxic coastal zones around the world, concentrated in coastal areas in Western Europe, the Eastern and Southern coasts of the US, and East Asia, particularly in Japan. In the ocean, there are frequent red tide algae blooms that kill fish and marine mammals and cause respiratory problems in humans and some domestic animals when the blooms reach close to shore.

In addition to land runoff, atmospheric anthropogenic fixed nitrogen can enter the open ocean. A study in 2008 found that this could account for around one third of the ocean's external (non-recycled) nitrogen supply and up to three per cent of the annual new marine biological production. It has been suggested that accumulating reactive nitrogen in the environment may have consequences as serious as putting carbon dioxide in the atmosphere.

## Acidification

The oceans are normally a natural carbon sink, absorbing carbon dioxide from the atmosphere. Because the levels of atmospheric carbon dioxide are increasing, the oceans are becoming more acidic. The potential consequences of ocean acidification are not fully understood, but there are concerns that structures made of calcium carbonate may become vulnerable to dissolution, affecting corals and the ability of shellfish to form shells.

A report from NOAA scientists published in the journal Science in May 2008 found that large amounts of relatively acidified water are upwelling to within four miles of the Pacific continental shelf area of North America. This area is a critical zone where most local marine life lives or is born. While the paper dealt only with areas from Vancouver to northern California, other continental shelf areas may be experiencing similar effects.

## Effects of Fishing

## Habitat Destruction

Fishing nets that have been left or lost in the ocean by fishermen are called ghost nets, and can entangle fish, dolphins, sea turtles, sharks, dugongs, crocodiles, seabirds, crabs, and other creatures. Acting as designed, these nets restrict movement, causing starvation, laceration and infection, and—in those that need to return to the surface to breathe—suffocation.

## Overfishing

Some specific examples of overfishing.

- On the east coast of the United States, the availability of bay scallops has been greatly diminished by the overfishing of sharks in the area. A variety of sharks have, until recently, fed on rays, which are a main predator of bay scallops. With the shark population reduced, in some places almost totally, the rays have been free to dine on scallops to the point of greatly decreasing their numbers.

- Chesapeake Bay's once-flourishing oyster populations historically filtered the estuary's entire water volume of excess nutrients every three or four days. Today that process takes almost a year, and sediment, nutrients, and algae can cause problems in local waters. Oysters filter these pollutants, and either eat them or shape them into small packets that are deposited on the bottom where they are harmless.

- The Australian government alleged in 2006 that Japan illegally overfished southern bluefin tuna by taking 12,000 to 20,000 tonnes per year instead of their agreed 6,000 tonnes; the value of such overfishing would be as much as US$2 billion. Such overfishing has resulted in severe damage to stocks. "Japan's huge appetite for tuna will take the most sought-after stocks to the brink of commercial extinction unless fisheries agree on more rigid quotas" stated the WWF. Japan disputes this figure, but acknowledges that some overfishing has occurred in the past.

- Jackson, Jeremy B C et al. (2001) *Historical overfishing and the recent collapse of coastal ecosystems* Science 293:629-638.

## Loss of Biodiversity

Each species in an ecosystem is affected by the other species in that ecosystem. There are very few single prey-single predator relationships. Most prey are consumed by more than one predator, and most predators have more than one prey. Their relationships are also influenced by other environmental factors. In most cases, if one species is removed from an ecosystem, other species will most likely be affected, up to the point of extinction.

Species biodiversity is a major contributor to the stability of ecosystems. When an organism exploits a wide range of resources, a decrease in biodiversity is less likely to have an impact. However, for an organism which exploit only limited resources, a decrease in biodiversity is more likely to have a strong effect.

Reduction of habitat, hunting and fishing of some species to extinction or near extinction, and pollution tend to tip the balance of biodiversity.

## Threatened Species

The global standard for recording threatened marine species is the IUCN Red List of Threatened Species. This list is the foundation for marine conservation priorities worldwide. A species is listed in the threatened category if it is considered to be critically endangered, endangered, or vulnerable. Other categories are near threatened and data deficient.

## Marine

Many marine species are under increasing risk of extinction and marine biodiversity is undergoing potentially irreversible loss due to threats such as overfishing, bycatch, climate change, invasive species and coastal development.

By 2008, the IUCN had assessed about 3,000 marine species. This includes assessments of known species of shark, ray, chimaera, reef-building coral, grouper, marine turtle, seabird, and marine mammal. Almost one-quarter (22%) of these groups have been listed as threatened.

| Group | Species | Threatened | Near threatened | Data deficient |
|---|---|---|---|---|
| Sharks, rays, and chimaeras | | 17% | 13% | 47% |
| Groupers | | 12% | 14% | 30% |
| Reef-building corals | 845 | 27% | 20% | 17% |
| Marine mammals | | 25% | | |
| Seabirds | | 27% | | |
| Marine turtles | 7 | 86% | | |

- Sharks, rays, and chimaeras: are deep water pelagic species, which makes them difficult to study in the wild. Not a lot is known about their ecology and population status. Much of what is currently known is from their capture in nets from both targeted and accidental catch. Many of these slow growing species are not recovering from overfishing by shark fisheries around the world.

- Groupers: Major threats are overfishing, particularly the uncontrolled fishing of small juveniles and spawning adults.

- Coral reefs: The primary threats to corals are bleaching and disease which has been linked to an increase in sea temperatures. Other threats include coastal development, coral extraction, sedimentation and pollution. The coral triangle (Indo-Malay-Philippine archipelago) region has the highest number of reef-building coral species in threatened category as well as the highest coral species diversity. The loss of coral reef ecosystems will have devastating effects on many marine species, as well as on people that depend on reef resources for their livelihoods.

- Marine mammals: include whales, dolphins, porpoises, seals, sea lions, walruses, sea otter, marine otter, manatees, dugong and the polar bear. Major threats include entanglement in ghost nets, targeted harvesting, noise pollution from military and seismic sonar, and boat strikes. Other threats are water pollution, habitat loss from coastal development, loss of food sources due to the collapse of fisheries, and climate change.

- Seabirds: Major threats include longline fisheries and gillnets, oil spills, and predation by rodents and cats in their breeding grounds. Other threats are habitat loss and degradation from coastal development, logging and pollution.

- Marine turtles: Marine turtles lay their eggs on beaches, and are subject to threats such as coastal development, sand mining, and predators, including humans who collect their eggs for food in many parts of the world. At sea, marine turtles can be targeted by small scale subsistence fisheries, or become bycatch during longline and trawling activities, or become entangled in ghost nets or struck by boats.

An ambitious project, called the Global Marine Species Assessment, is under way to make IUCN Red List assessments for another 17,000 marine species by 2012. Groups targeted include the approximately 15,000 known marine fishes, and important habitat-forming primary producers such mangroves, seagrasses, certain seaweeds and the remaining corals; and important invertebrate groups including molluscs and echinoderms.

## Freshwater

Freshwater fisheries have a disproportionately high diversity of species compared to other ecosystems. Although freshwater habitats cover less than 1% of the world's surface, they provide a home for over 25% of known vertebrates, more than 126,000 known animal species, about 24,800 species of freshwater fish, molluscs, crabs and dragonflies, and about 2,600 macrophytes. Continuing industrial and agricultural developments place huge strain on these freshwater systems. Waters are polluted or extracted at high levels, wetlands are drained, rivers channelled, forests deforestated leading to sedimentation, invasive species are introduced, and over-harvesting occurs.

In the 2008 IUCN Red List, about 6,000 or 22% of the known freshwater species have been assessed at a global scale, leaving about 21,000 species still to be assessed. This makes clear that, worldwide, freshwater species are highly threatened, possibly more so than species in marine fisheries. However, a significant proportion of freshwater species are listed as data deficient, and more field surveys are needed.

## Fisheries Management

A recent paper published by the National Academy of Sciences of the USA warns that: "Synergistic effects of habitat destruction, overfishing, introduced species, warming, acidification, toxins, and massive runoff of nutrients are transforming once complex ecosystems like coral reefs and kelp forests into monotonous level bottoms, transforming clear and productive coastal seas into anoxic dead zones, and transforming complex food webs topped by big animals into simplified, microbially dominated ecosystems with boom and bust cycles of toxic dinoflagellate blooms, jellyfish, and disease".

# References

- Norse, Elliott A. and Crowder, Larry B. (Eds.) (2005) Marine Conservation Biology: The Science of Maintaining the Sea's Biodiversity, Island Press. ISBN 978-1-55963-662-9

- McLeod, Karen and Leslie. Heather (Eds.) (2009) Ecosystem-Based Management for the Oceans Island Press. ISBN 978-1-59726-155-5

- Berkes F, Mahon R, McConney P, Pollnac R and Pomeroy R (2001) Managing Small-Scale Fisheries: Alternative Directions and Methods IDRC,. ISBN 978-0-88936-943-6

- Mann, Kenneth and Lazier, John (3rd Ed. 2005) Dynamics of Marine Ecosystems: Biological-Physical Interactions in the Oceans Wiley-Blackwell. ISBN 978-1-4051-1118-8

- Norse EA and Crowder LB (Eds) (2005) Marine conservation biology: the science of maintaining the sea's biodiversity Island Press. ISBN 978-1-55963-662-9

- De Young, Cassandra (2007) Review of the state of world marine capture fisheries management FAO, Fisheries Technical Paper 488, Rome. ISBN 978-92-5-105875-6.

# An Overview of Different Fish Diseases and Parasites

Aquaculture management includes the growth and well-being of the marine species that are being cultivated. The knowledge of diseases that afflict fish and other aquatic animals are also very necessary. This chapter touches upon the various diseases and parasites that are found among aquatic animals.

## Fish Diseases and Parasites

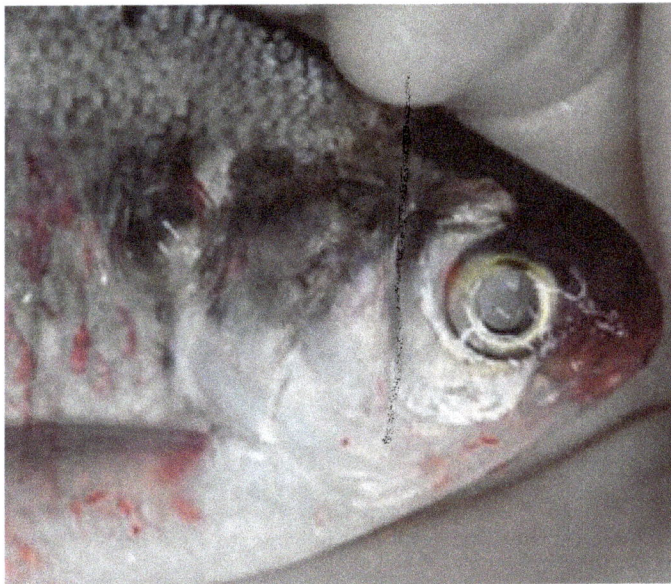

This gizzard shad has VHS, a deadly infectious disease which causes bleeding. It afflicts over 50 species of freshwater and marine fish in the northern hemisphere.

Like humans and other animals, fish suffer from diseases and parasites. Fish defences against disease are specific and non-specific. Non-specific defences include skin and scales, as well as the mucus layer secreted by the epidermis that traps microorganisms and inhibits their growth. If pathogens breach these defences, fish can develop inflammatory responses that increase the flow of blood to infected areas and deliver white blood cells that attempt to destroy the pathogens.

Specific defences are specialised responses to particular pathogens recognised by the fish's body, that is adaptive immune responses. In recent years, vaccines have become widely used in aquaculture and ornamental fish, for example vaccines for furunculosis in farmed salmon and koi herpes virus in koi.

This flatfish *Limanda limanda* has an outgrowth called a xenoma. It is caused
by a microsporidian fungal parasite in its intestines.

Some commercially important fish diseases are VHS, ich and whirling disease.

## Disease

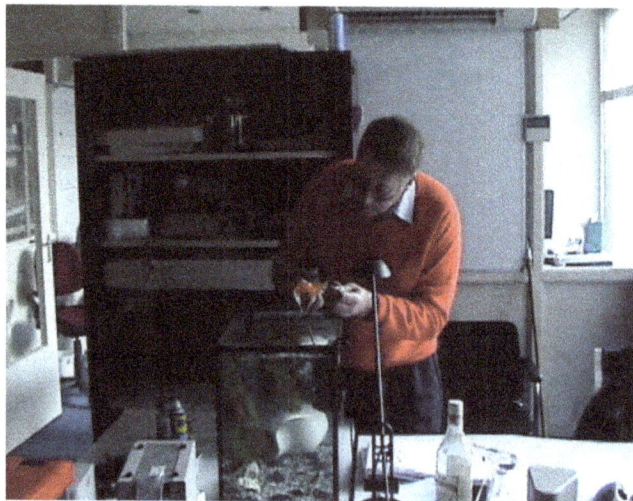

A veterinarian gives an injection to a goldfish

All fish carry pathogens and parasites. Usually this is at some cost to the fish. If the cost is sufficiently high, then the impacts can be characterised as a disease. However disease in fish is not understood well. What is known about fish disease often relates to aquaria fish, and more recently, to farmed fish.

Disease is a prime agent affecting fish mortality, especially when fish are young. Fish can limit the impacts of pathogens and parasites with behavioural or biochemical means, and such fish have reproductive advantages. Interacting factors result in low grade infection becoming fatal diseases. In particular, things that causes stress, such as natural droughts or pollution or predators, can precipitate outbreak of disease.

Disease can also be particularly problematic when pathogens and parasites carried by introduced species affect native species. An introduced species may find invading easier if potential predators and competitors have been decimated by disease.

Pathogens which can cause fish diseases comprise:

- viral infections

- bacterial infections, such as *Pseudomonas fluorescens* leading to fin rot and fish dropsy

- fungal infections

- water mould infections, such as *Saprolegnia* sp.

- metazoan parasites, such as copepods

- unicellular parasites, such as *Ichthyophthirius multifiliis* leading to ich

- Certain parasites like Helminths for example *Eustrongylides*

## Parasites

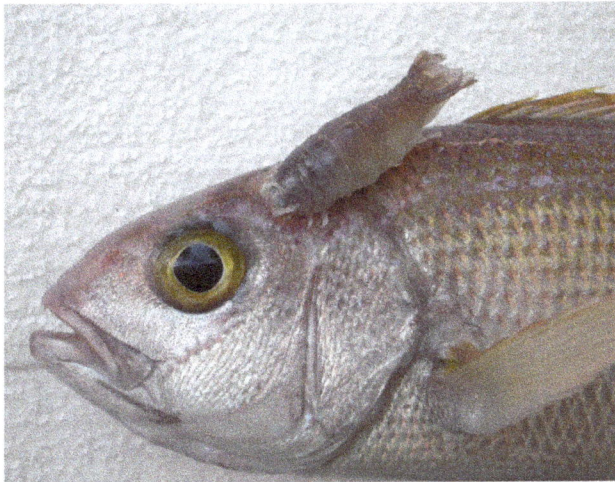

The isopod Anilocra gigantea parasitising the snapper *Pristipomoides filamentosus*

Parasites in fish are a common natural occurrence. Parasites can provide information about host population ecology. In fisheries biology, for example, parasite communities can be used to distinguish distinct populations of the same fish species co-inhabiting a region. Additionally, parasites possess a variety of specialized traits and life-history strategies that enable them to colonize hosts. Understanding these aspects of parasite ecology, of interest in their own right, can illuminate parasite-avoidance strategies employed by hosts.

Usually parasites (and pathogens) need to avoid killing their hosts, since extinct hosts can mean extinct parasites. Evolutionary constraints may operate so parasites avoid killing their hosts, or the natural variability in host defensive strategies may suffice to keep host populations viable. Parasite infections can impair the courtship dance of male threespine sticklebacks. When that happens, the females reject them, suggesting a strong mechanism for the selection of parasite resistance."

However, not all parasites want to keep their hosts alive, and there are parasites with multistage life cycles who go to some trouble to kill their host. For example, some tapeworms make some fish behave in such a way that a predatory bird can catch it. The predatory bird is the next host for the parasite in the next stage of its life cycle. Specifically, the tapeworm *Schistocephalus solidus* turns infected threespine stickleback white, and then makes them more buoyant so that they splash along at the surface of the water, becoming easy to see and easy to catch for a passing bird.

*Cymothoa exigua* is a parasitic crustacean which enters a fish through its gills and destroys the fish's tongue.

Parasites can be internal (endoparasites) or external (ectoparasites). Some internal fish parasites are spectacular, such as the philometrid nematode *Philometra fasciati* which is parasitic in the ovary of female Blacktip grouper; the adult female parasite is a red worm which can reach up to 40 centimetres in length, for a diameter of only 1.6 millimetre; the males are tiny. Other internal parasites are found living inside fish gills, include encysted adult didymozoid trematodes, a few trichosomoidid nematodes of the genus *Huffmanela*, including *Huffmanela ossicola* which lives within the gill bone, and the encysted parasitic turbellarian *Paravortex*. Various protists and Myxosporea are also parasitic on gills, where they form cysts.

Fish gills are also the preferred habitat of many external parasites, attached to the gill but living out of it. The most common are monogeneans and certain groups of parasitic copepods, which can be extremely numerous. Other external parasites found on gills are leeches and, in seawater, larvae of gnathiid isopods. Isopod fish parasites are mostly external and feed on blood. The larvae of the Gnathiidae family and adult cymothoidids have piercing and sucking mouthparts and clawed limbs adapted for clinging onto their hosts. *Cymothoa exigua* is a parasite of various marine fish. It causes the tongue of the fish to atrophy and takes its place in what is believed to be the first instance discovered of a parasite functionally replacing a host structure in animals.

Other parasitic disorders, include *Gyrodactylus salaris*, *Ichthyophthirius multifiliis*, cryptocaryon, velvet disease, *Brooklynella hostilis*, Hole in the head, *Glugea*, *Ceratomyxa shasta*, *Kudoa thyrsites*, *Tetracapsuloides bryosalmonae*, *Cymothoa exigua*, leeches, nematode, flukes, carp lice and salmon lice.

Although parasites are generally considered to be harmful, the eradication of all parasites would not necessarily be beneficial. Parasites account for as much as or more than half of life's diversity; they perform an important ecological role (by weakening prey) that ecosystems would take some time to adapt to; and without parasites organisms may eventually tend to asexual reproduction, diminishing the diversity of sexually dimorphic traits. Parasites provide an opportunity for the transfer of genetic material between species. On rare, but significant, occasions this may facilitate evolutionary changes that would not otherwise occur, or that would otherwise take even longer.

Below are some life cycles of fish parasites:

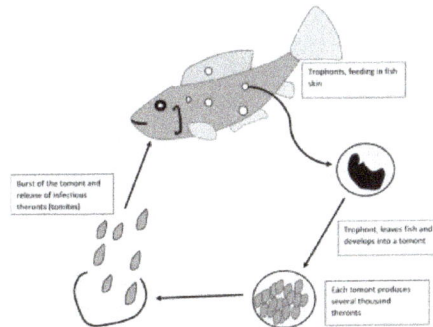

Life cycle of the fish parasite *Ichthyophthirius multifiliis*, commonly called ich

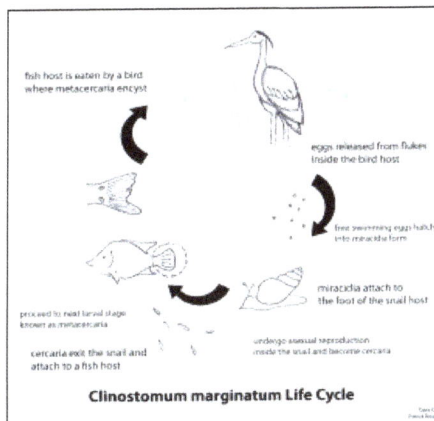

Life cycle of the parasitic fluke *Clinostomum marginatum*, commonly called the yellow grub

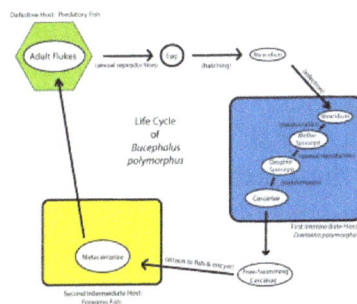

Life cycle of the digenean *Bucephalus polymorphus*

## Cleaner Fish

Some fish take advantage of cleaner fish for the removal of external parasites. The best known of these are the Bluestreak cleaner wrasses of the genus *Labroides* found on coral reefs in the Indian Ocean and Pacific Ocean. These small fish maintain so-called "cleaning stations" where other fish, known as hosts, will congregate and perform specific movements to attract the attention of the cleaner fish. Cleaning behaviours have been observed in a number of other fish groups, including an interesting case between two cichlids of the same genus, *Etroplus maculatus*, the cleaner fish, and the much larger *Etroplus suratensis*, the host.

Two cleaner wrasses, *Labroides phthirophagus*, servicing a goatfish, *Mulloidichthys flavolineatus*

More than 40 species of parasites may reside on the skin and internally of the ocean sunfish, motivating the fish to seek relief in a number of ways. In temperate regions, drifting kelp fields harbour cleaner wrasses and other fish which remove parasites from the skin of visiting sunfish. In the tropics, the *mola* will solicit cleaner help from reef fishes. By basking on its side at the surface, the sunfish also allows seabirds to feed on parasites from their skin. Sunfish have been reported to breach more than ten feet above the surface, possibly as another effort to dislodge parasites on the body.

## Mass Die Offs

Some diseases result in mass die offs. One of the more bizarre and recently discovered diseases produces huge fish kills in shallow marine waters. It is caused by the ambush predator dinoflagellate *Pfiesteria piscicida*. When large numbers of fish, like shoaling forage fish, are in confined situations such as shallow bays, the excretions from the fish encourage this dinoflagellate, which is not normally toxic, to produce free-swimming zoospores. If the fish remain in the area, continuing to provide nourishment, then the zoospores start secreting a neurotoxin. This toxin results in the fish developing bleeding lesions, and their skin flakes off in the water. The dinoflagellates then eat the blood and flakes of tissue while the affected fish die. Fish kills by this dinoflagellate are common, and they may also have been responsible for kills in the past which were thought to have had other causes. Kills like these can be viewed as natural mechanisms for regulating the population of exceptionally abundant fish. The rate at which the kills occur increases as organically polluted land runoff increases.

## Wild Salmon

According to Canadian biologist Dorothy Kieser, protozoan parasite *Henneguya salminicola* is commonly found in the flesh of salmonids. It has been recorded in the field samples of salmon returning to the Queen Charlotte Islands. The fish responds by walling off the parasitic infection into a number of cysts that contain milky fluid. This fluid is an accumulation of a large number of parasites.

*Henneguya salminicola*, a parasite commonly found in the flesh of salmonids
on the West Coast of Canada. Coho salmon

*Henneguya* and other parasites in the myxosporean group have a complex lifecycle where the salmon is one of two hosts. The fish releases the spores after spawning. In the *Henneguya* case, the spores enter a second host, most likely an invertebrate, in the spawning stream. When juvenile salmon out-migrate to the Pacific Ocean, the second host releases a stage infective to salmon. The parasite is then carried in the salmon until the next spawning cycle. The myxosporean parasite that causes whirling disease in trout, has a similar lifecycle. However, as opposed to whirling disease, the *Henneguya* infestation does not appear to cause disease in the host salmon — even heavily infected fish tend to return to spawn successfully.

According to Dr. Kieser, a lot of work on *Henneguya salminicola* was done by scientists at the Pacific Biological Station in Nanaimo in the mid-1980s, in particular, an overview report which states that "the fish that have the longest fresh water residence time as juveniles have the most noticeable infections. Hence in order of prevalence coho are most infected followed by sockeye, chinook, chum and pink." As well, the report says that, at the time the studies were conducted, stocks from the middle and upper reaches of large river systems in British Columbia such as Fraser, Skeena, Nass and from mainland coastal streams in the southern half of B.C. "are more likely to have a low prevalence of infection." The report also states "It should be stressed that *Henneguya*, economically deleterious though it is, is harmless from the view of public health. It is strictly a fish parasite that cannot live in or affect warm blooded animals, including man".

According to Klaus Schallie, Molluscan Shellfish Program Specialist with the Canadian Food Inspection Agency, "*Henneguya salminicola* is found in southern B.C. also and in all species of salmon. I have previously examined smoked chum salmon sides that were riddled with cysts and some sockeye runs in Barkley Sound (southern B.C., west coast of Vancouver Island) are noted for their high incidence of infestation."

Sea lice, particularly *Lepeophtheirus salmonis* and a variety of *Caligus* species, including *Caligus clemensi* and *Caligus rogercresseyi*, can cause deadly infestations of both farm-grown and wild salmon. Sea lice are ectoparasites which feed on mucous, blood, and skin, and migrate and latch onto the skin of wild salmon during free-swimming, planktonic *naupli* and *copepodid* larval stages, which can persist for several days. Large numbers of highly populated, open-net salmon farms can create exceptionally large concentrations of sea lice; when exposed in river estuaries containing large numbers of open-net farms, many young wild salmon are infected, and do not survive as a result. Adult salmon may survive otherwise critical numbers of sea lice, but small, thin-skinned juvenile salmon migrating to sea are highly vulnerable. On the Pacific coast of Canada, the louse-induced mortality of pink salmon in some regions is commonly over 80%.

Sample of pink salmon infected with Henneguya salminicola, caught off the Queen Charlotte Islands, Western Canada in 2009

## Farmed Salmon

In 1972, Gyrodactylus salaris, also called salmon fluke, a monogenean parasite, spread from Norwegian hatcheries to wild salmon, and devastated some wild salmon populations.

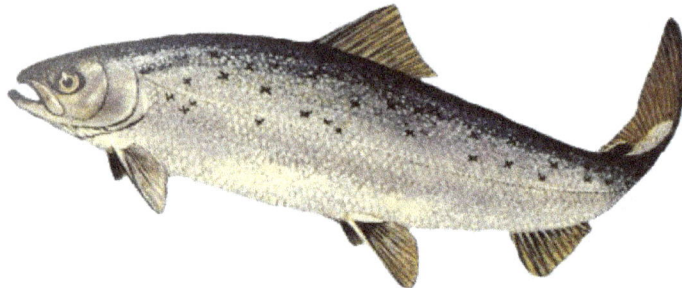

Atlantic salmon

In 1984, infectious salmon anemia (ISAv) was discovered in Norway in an Atlantic salmon hatchery. Eighty percent of the fish in the outbreak died. ISAv, a viral disease, is now a major threat to the viability of Atlantic salmon farming. It is now the first of the diseases classified on List One of the European Commission's fish health regime. Amongst other measures, this requires the total eradication of the entire fish stock should an outbreak of the disease be confirmed on any farm. ISAv seriously affects salmon farms in Chile, Norway, Scotland and Canada, causing major economic losses to infected farms. As the name implies, it causes severe anemia of infected fish. Unlike mammals, the red blood cells of fish have DNA, and can become infected with viruses. The fish develop pale gills, and may swim close to the water surface, gulping for air. However, the disease can also develop without the fish showing any external signs of illness, the fish maintain a normal appetite, and then they suddenly die. The disease can progress slowly throughout an infected farm and, in the worst cases, death rates may approach 100 percent. It is also a threat to the dwindling stocks of wild salmon. Management strategies include developing a vaccine and improving genetic resistance to the disease.

In the wild, diseases and parasites are normally at low levels, and kept in check by natural predation on weakened individuals. In crowded net pens they can become epidemics. Diseases and parasites also transfer from farmed to wild salmon populations. A recent study in British Columbia links the spread of parasitic sea lice from river salmon farms to wild pink salmon in the same river." The European Commission (2002) concluded "The reduction of wild salmonid abundance is also linked to other factors but there is more and more scientific evidence establishing a direct link between the number of lice-infested wild fish and the presence of cages in the same estuary." It is reported that wild salmon on the west coast of Canada are being driven to extinction by sea lice from nearby salmon farms. Antibiotics and pesticides are often used to control the diseases and parasites.

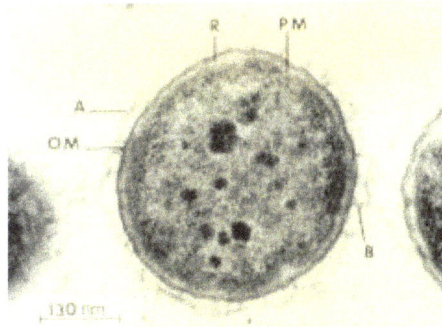

*Aeromonas salmonicida*, a Gram-negative bacteria, causes the disease furunculosis in marine and freshwater fish.

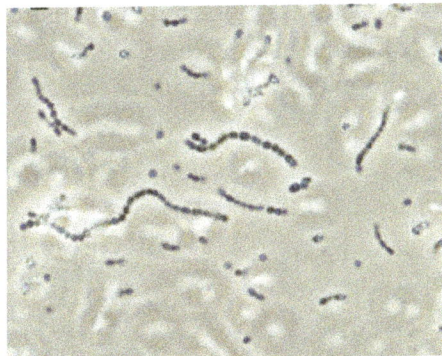

*Streptococcus iniae*, a Gram-positive, sphere-shaped bacteria caused losses in farmed marine and freshwater finfish of US$100 million in 1997.

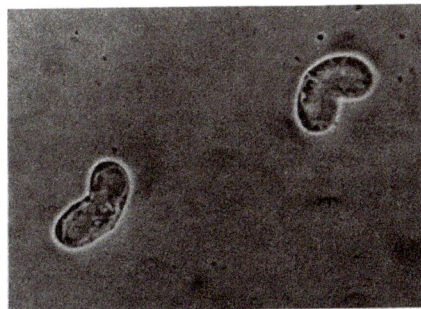

*Ceratomyxa shasta*, another myxosporean parasite, infects salmonid fish on the Pacific coast of North America.

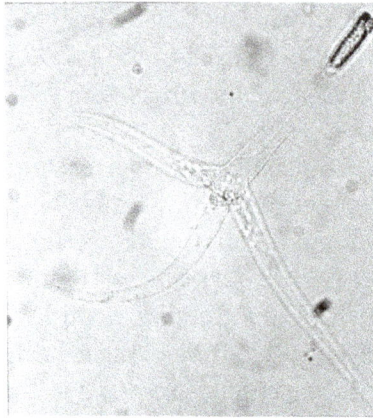

*Myxobolus cerebralis*, a myxosporean parasite, causes *whirling disease* in farmed salmon and trout and also in wild fish populations.

## Coral Reef Fish

Coral reef fish are characterized by high biodiversity. As a consequence parasites of coral reef fish show tremendous variety. Parasites of coral reef fish include nematodes, Platyhelminthes (cestodes, digeneans, and monogeneans), leeches, parasitic crustaceans such as isopods and copepods, and various microorganisms such as myxosporidia and microsporidia. Some of these fish parasites have heteroxenous life cycles (i.e. they have several hosts) among which sharks (certain cestodes) or molluscs (digeneans). The high biodiversity of coral reefs increases the complexity of the interactions between parasites and their various and numerous hosts. Numerical estimates of parasite biodiversity have shown that certain coral fish species have up to 30 species of parasites. The mean number of parasites per fish species is about ten. This has a consequence in term of co-extinction. Results obtained for the coral reef fish of New Caledonia suggest that extinction of a coral reef fish species of average size would eventually result in the co-extinction of at least ten species of parasites.

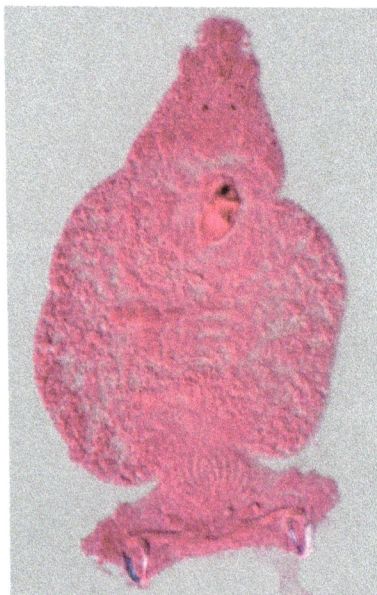

Monogenean parasite on the gill of a grouper

## Aquarium Fish

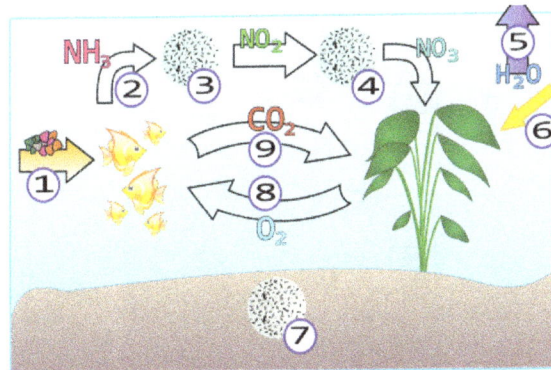

Nitrogen cycle in a common aquarium.

In most aquarium tanks, the fish are at high concentrations and the volume of water is limited. This means that communicable diseases can spread rapidly to most or all fish in a tank. An improper nitrogen cycle, inappropriate aquarium plants and potentially harmful freshwater invertebrates can directly harm or add to the stresses on ornamental fish in a tank. Despite this, many diseases in captive fish can be avoided or prevented through proper water conditions and a well-adjusted ecosystem within the tank. Ammonia poisoning is a common disease in new aquariums, especially when immediately stocked to full capacity.

Ornamental fish kept in aquariums are susceptible to numerous diseases.

Due to their generally small size and the low cost of replacing diseased or dead aquarium fish, the cost of testing and treating diseases is often seen as more trouble than the value of the fish.

Goldfish with dropsy

Columnaris in the gill of a chinook salmon

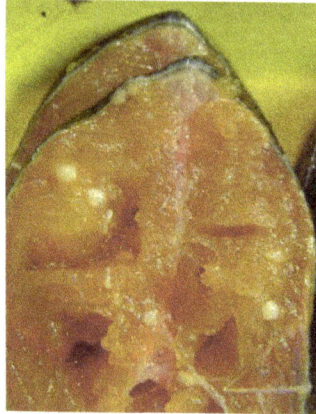

The parasite *Henneguya zschokkei* in salmon beard

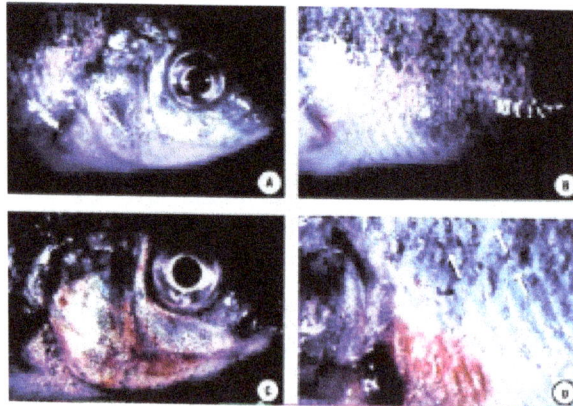

Skin ulcers in tilapia exposed to Pfiesteria shumwayae

## Immune System

Immune organs vary by type of fish. In the jawless fish (lampreys and hagfish), true lymphoid organs are absent. These fish rely on regions of lymphoid tissue within other organs to produce immune cells. For example, erythrocytes, macrophages and plasma cells are produced in the anterior kidney (or pronephros) and some areas of the gut (where granulocytes mature.) They resemble primitive bone marrow in hagfish. Cartilaginous fish (sharks and rays) have a more advanced immune system. They have three specialized organs that are unique to chondrichthyes; the epigonal organs (lymphoid tissue similar to mammalian bone) that surround the gonads, the Leydig's organ within the walls of their esophagus, and a spiral valve in their intestine. These organs house typical immune cells (granulocytes, lymphocytes and plasma cells). They also possess an identifiable thymus and a well-developed spleen (their most important immune organ) where various lymphocytes, plasma cells and macrophages develop and are stored. Chondrostean fish (sturgeons, paddlefish and bichirs) possess a major site for the production of granulocytes within a mass that is associated with the meninges (membranes surrounding the central nervous system.) Their heart is frequently covered with tissue that contains lymphocytes, reticular cells and a small number of macrophages. The chondrostean kidney is an important hemopoietic organ; where erythrocytes, granulocytes, lymphocytes and macrophages develop.

Like chondrostean fish, the major immune tissues of bony fish (or teleostei) include the kidney (especially the anterior kidney), which houses many different immune cells. In addition, teleost fish possess a thymus, spleen and scattered immune areas within mucosal tissues (e.g. in the skin, gills, gut and gonads). Much like the mammalian immune system, teleost erythrocytes, neutrophils and granulocytes are believed to reside in the spleen whereas lymphocytes are the major cell type found in the thymus. In 2006, a lymphatic system similar to that in mammals was described in one species of teleost fish, the zebrafish. Although not confirmed as yet, this system presumably will be where naive (unstimulated) T cells accumulate while waiting to encounter an antigen.

## Spreading Disease and Parasites

The capture, transportation and culture of bait fish can spread damaging organisms between ecosystems, endangering them. In 2007, several American states, including Michigan, enacted regulations designed to slow the spread of fish diseases, including viral hemorrhagic septicemia, by bait fish. Because of the risk of transmitting *Myxobolus cerebralis* (whirling disease), trout and salmon should not be used as bait. Anglers may increase the possibility of contamination by emptying bait buckets into fishing venues and collecting or using bait improperly. The transportation of fish from one location to another can break the law and cause the introduction of fish and parasites alien to the ecosystem.

## Eating Raw Fish

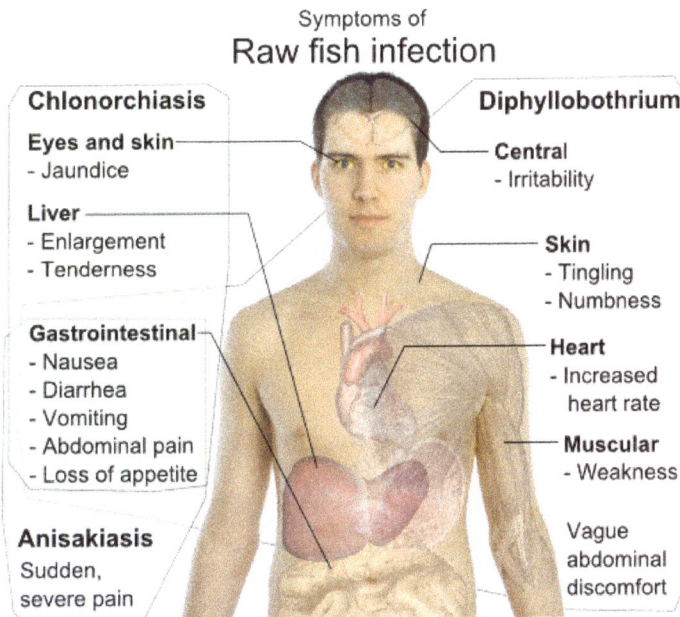

Differential symptoms of parasite infection by raw fish: Clonorchis sinensis (a trematode/fluke), Anisakis (a nematode/roundworm) and Diphyllobothrium a (cestode/tapeworm), all have gastrointestinal, but otherwise distinct, symptoms.

Though not a health concern in thoroughly cooked fish, parasites are a concern when human consumers eat raw or lightly preserved fish such as sashimi, sushi, ceviche, and gravlax. The popularity of such raw fish dishes makes it important for consumers to be aware of this risk. Raw fish should be frozen to an internal temperature of −20 °C (−4 °F) for at least 7 days to kill parasites. It is important to be aware that home freezers may not be cold enough to kill parasites.

Traditionally, fish that live all or part of their lives in fresh water were considered unsuitable for sashimi due to the possibility of parasites. Parasitic infections from fresh-water fish are a serious problem in some parts of the world, particularly Southeast Asia. Fish that spend part of their life cycle in salt water, like salmon, can also be a problem. A study in Seattle, Washington showed that 100% of wild salmon had roundworm larvae capable of infecting people. In the same study farm raised salmon did not have any roundworm larvae.

Parasite infection by raw fish is rare in the developed world (fewer than 40 cases per year in the U.S.), and involves mainly three kinds of parasites: Clonorchis sinensis (a trematode/fluke), Anisakis (a nematode/roundworm) and Diphyllobothrium (a cestode/tapeworm). Infection by the fish tapeworm *Diphyllobothrium latum* is seen in countries where people eat raw or undercooked fish, such as some countries in Asia, Eastern Europe, Scandinavia, Africa, and North and South America. Infection risk of anisakis is particularly higher in fishes which may live in a river such as salmon (*shake*) in Salmonidae, mackerel (*saba*). Such parasite infections can generally be avoided by boiling, burning, preserving in salt or vinegar, or freezing overnight. Even Japanese people never eat raw salmon or ikura (salmon roe), and even if they seem raw, these foods are not raw but are frozen overnight to prevent infections from parasites, particularly anisakis.

Below are some life cycles of fish parasites that can infect humans:

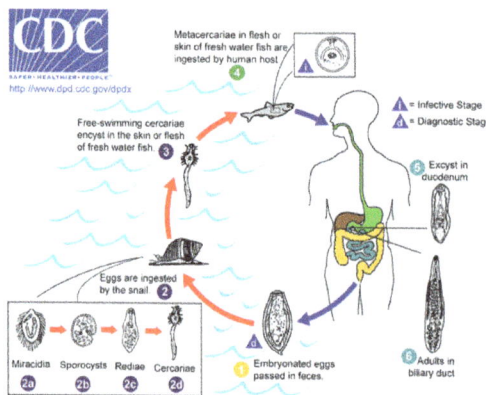

Life cycle of the liver fluke *Clonorchis sinensis*

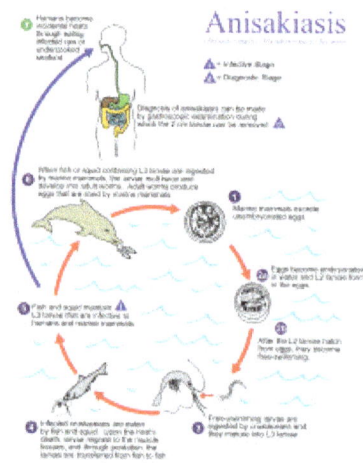

Life cycle of the parasitic *Anisakis* worm

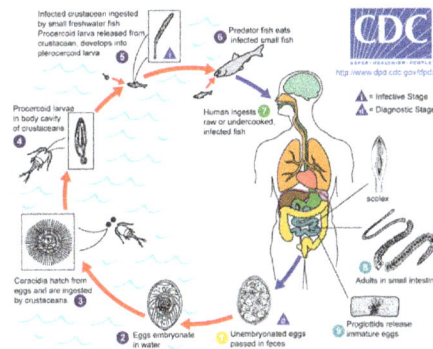

Life cycle of the fish tapeworm *Diphyllobothrium latum*

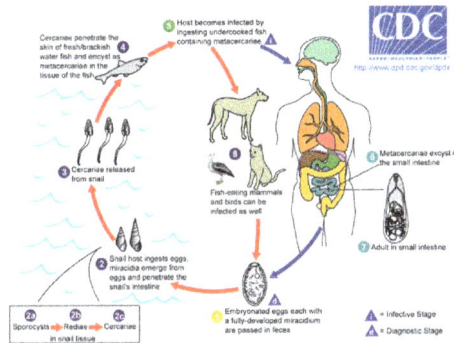

Life cycle of the digenean *Metagonimus*, an intestinal fluke

# Diseases and Parasites in Salmon

Sample of pink salmon infected with Henneguya salminicola, caught off the Queen Charlotte Islands, Western Canada in 2009

This article is about diseases and parasites in salmon, trout and other salmon-like fishes of the Salmonidae family.

According to Canadian biologist Dorothy Kieser, the myxozoan parasite *Henneguya salminicola* is commonly found in the flesh of salmonids. It has been recorded in the field samples of salmon returning to the Queen Charlotte Islands. The fish responds by walling off the parasitic infection into a number of cysts that contain milky fluid. This fluid is an accumulation of a large number of parasites.

*Henneguya* and other parasites in the myxosporean group have a complex life cycle, where the salmon is one of two hosts. The fish releases the spores after spawning. In the *Henneguya* case, the spores enter a second host, most likely an invertebrate, in the spawning stream. When juvenile salmon migrate to the Pacific Ocean, the second host releases a stage infective to salmon. The parasite is then carried in the salmon until the next spawning cycle. The myxosporean parasite that causes whirling disease in trout, has a similar life cycle. However, as opposed to whirling disease, the *Henneguya* infestation does not appear to cause significant incapacitation of the host salmon — even heavily infected fish tend to return to spawn successfully.

According to Dr. Kieser, a lot of work on *Henneguya salminicola* was done by scientists at the Pacific Biological Station in Nanaimo in the mid-1980s, in particular, an overview report which states that "the fish that have the longest fresh water residence time as juveniles have the most noticeable infections. Hence in order of prevalence coho are most infected followed by sockeye, chinook, chum and pink." As well, the report says that, at the time the studies were conducted, stocks from the middle and upper reaches of large river systems in British Columbia such as Fraser, Skeena, Nass and from mainland coastal streams in the southern half of B.C. "are more likely to have a low prevalence of infection." The report also states "It should be stressed that *Henneguya*, economically deleterious though it is, is harmless from the view of public health. It is strictly a fish parasite that cannot live in or affect warm blooded animals, including man".

According to Klaus Schallie, Molluscan Shellfish Program Specialist with the Canadian Food Inspection Agency, "*Henneguya salminicola* is found in southern B.C. also and in all species of salmon. I have previously examined smoked chum salmon sides that were riddled with cysts and some sockeye runs in Barkley Sound (southern B.C., west coast of Vancouver Island) are noted for their high incidence of infestation."

Sea lice, particularly *Lepeophtheirus salmonis* and various *Caligus* species, including *C. clemensi* and *C. rogercresseyi*, can cause deadly infestations of both farm-grown and wild salmon. Sea lice are ectoparasites which feed on mucus, blood, and skin, and migrate and latch onto the skin of wild salmon during free-swimming, planktonic nauplii and copepodid larval stages, which can persist for several days. Large numbers of highly populated, open-net salmon farms can create exceptionally large concentrations of sea lice; when exposed in river estuaries containing large numbers of open-net farms, many young wild salmon are infected, and do not survive as a result. Adult salmon may survive otherwise critical numbers of sea lice, but small, thin-skinned juvenile salmon migrating to sea are highly vulnerable. On the Pacific coast of Canada, the louse-induced mortality of pink salmon in some regions is commonly over 80%.

## Some Background

In 1984, infectious salmon anemia (ISAv) was discovered in Norway in an Atlantic salmon hatchery. Eighty percent of the fish in the outbreak died. ISAv, a viral disease, is now a major threat to the viability of Atlantic salmon farming. It is now the first of the diseases classified on List One of the European Commission's fish health regime. Amongst other measures, this requires the total eradication of the entire fish stock should an outbreak of the disease be confirmed on any farm. ISAv seriously affects salmon farms in Chile, Norway, Scotland and Canada, causing major economic losses to infected farms. As the name implies, it causes severe anemia of infected fish. Unlike mammals, the red blood cells of fish have DNA, and can become infected with viruses. The fish develop pale gills, and may swim close to the water surface, gulping for air. However, the disease can also develop without the fish showing any external signs of illness, the fish maintain a normal appetite, and then they suddenly die. The disease can progress slowly throughout an infected farm and, in the worst cases, death rates may approach 100 percent. It is also a threat to the dwindling stocks of wild salmon. Management strategies include developing a vaccine and improving genetic resistance to the disease.

Sea lice on the back of an Atlantic salmon

In 1972, Gyrodactylus, a monogenean parasite, spread from Norwegian hatcheries to wild salmon, and devastated some wild salmon populations.

In the wild, diseases and parasites are normally at low levels, and kept in check by natural predation on weakened individuals. In crowded net pens they can become epidemics. Diseases and parasites also transfer from farmed to wild salmon populations. A recent study in British Columbia links the spread of parasitic sea lice from river salmon farms to wild pink salmon in the same river. The European Commission (2002) concluded "The reduction of wild salmonid abundance is also linked to other factors but there is more and more scientific evidence establishing a direct link between the number of lice-infested wild fish and the presence of cages in the same estuary." It is reported that wild salmon on the west coast of Canada are being driven to extinction by sea lice from nearby salmon farms. These predictions have been disputed by other scientists and recent harvests have indicated that the predictions were in error. Antibiotics and pesticides are often used to control the diseases and parasites.

## Wild Salmon

According to Canadian biologist Dorothy Kieser, protozoan parasite *Henneguya salminicola* is commonly found in the flesh of salmonids. It has been recorded in the field samples of salmon returning to the Queen Charlotte Islands. The fish responds by walling off the parasitic infection into a number of cysts that contain milky fluid. This fluid is an accumulation of a large number of parasites.

*Henneguya* and other parasites in the myxosporean group have a complex lifecycle where the salmon is one of two hosts. The fish releases the spores after spawning. In the *Henneguya* case, the spores enter a second host, most likely an invertebrate, in the spawning stream. When juvenile salmon out-migrate to the Pacific Ocean, the second host releases a stage infective to salmon. The parasite is then carried in the salmon until the next spawning cycle. The myxosporean parasite that causes whirling disease in trout, has a similar lifecycle. However, as opposed to whirling disease, the *Henneguya* infestation does not appear to cause disease in the host salmon — even heavily infected fish tend to return to spawn successfully.

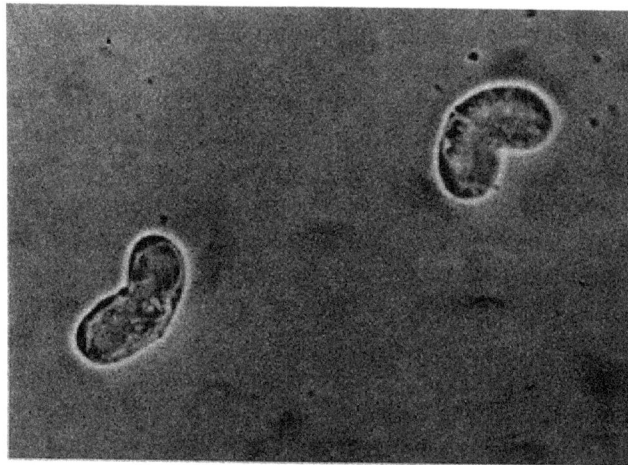

The myxosporean parasite *Ceratomyxa shasta* infects salmonid fish on the Pacific coast of North America

According to Dr. Kieser, a lot of work on *Henneguya salminicola* was done by scientists at the Pacific Biological Station in Nanaimo in the mid-1980s, in particular, an overview report which states that "the fish that have the longest fresh water residence time as juveniles have the most noticeable infections. Hence in order of prevalence coho are most infected followed by sockeye, chinook, chum and pink." As well, the report says that, at the time the studies were conducted, stocks from the middle and upper reaches of large river systems in British Columbia such as Fraser, Skeena, Nass and from mainland coastal streams in the southern half of B.C. "are more likely to have a low prevalence of infection." The report also states "It should be stressed that *Henneguya*, economically deleterious though it is, is harmless from the view of public health. It is strictly a fish parasite that cannot live in or affect warm blooded animals, including man".

According to Klaus Schallie, Molluscan Shellfish Program Specialist with the Canadian Food Inspection Agency, "*Henneguya salminicola* is found in southern B.C. also and in all species of salmon. I have previously examined smoked chum salmon sides that were riddled with cysts and some sockeye runs in Barkley Sound (southern B.C., west coast of Vancouver Island) are noted for their high incidence of infestation."

Sea lice, particularly *Lepeophtheirus salmonis* and a variety of *Caligus* species, including *Caligus clemensi* and *Caligus rogercresseyi*, can cause deadly infestations of both farm-grown and wild salmon. Sea lice are ectoparasites which feed on mucous, blood, and skin, and migrate and latch onto the skin of wild salmon during free-swimming, planktonic *naupli* and *copepodid* larval stages, which can persist for several days. Large numbers of highly populated, open-net salmon farms can create exceptionally large concentrations of sea lice; when exposed in river estuaries containing large numbers of open-net farms, many young wild salmon are infected, and do not survive as a result. Adult salmon may survive otherwise critical numbers of sea lice, but small, thin-skinned juvenile salmon migrating to sea are highly vulnerable. On the Pacific coast of Canada, the louse-induced mortality of pink salmon in some regions is commonly over 80%.

## Farmed Salmon

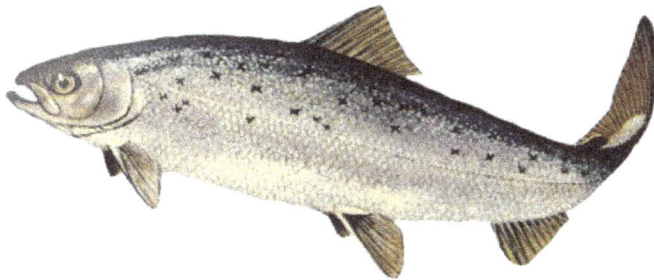

Atlantic salmon

In 1972, Gyrodactylus, a monogenean parasite, spread from Norwegian hatcheries to wild salmon, and devastated some wild salmon populations. In 1984, infectious salmon anemia (ISAv) was discovered in Norway in an Atlantic salmon hatchery. Eighty percent of the fish in the outbreak died. ISAv, a viral disease, is now a major threat to the viability of Atlantic salmon farming. It is now the first of the diseases classified on List One of the European Commission's fish health regime. Amongst other measures, this requires the total eradication of the entire fish stock should an outbreak of the disease be confirmed on any farm. ISAv seriously affects salmon farms in Chile, Norway, Scotland and Canada, causing major economic losses to infected farms. As the name implies, it causes severe anemia of infected fish. Unlike mammals, the red blood cells of fish have DNA, and can become infected with viruses. The fish develop pale gills, and may swim close to the water surface, gulping for air. However, the disease can also develop without the fish showing any external signs of illness, the fish maintain a normal appetite, and then they suddenly die. The disease can progress slowly throughout an infected farm and, in the worst cases, death rates may approach 100 percent. It is also a threat to the dwindling stocks of wild salmon. Management strategies include developing a vaccine and improving genetic resistance to the disease.

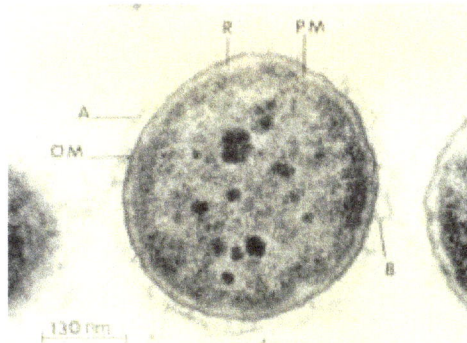

*Aeromonas salmonicida*, a Gram-negative bacteria, causes the disease furunculosis in marine and freshwater fish.

In the wild, diseases and parasites are normally at low levels, and kept in check by natural predation on weakened individuals. In crowded net pens they can become epidemics. Diseases and parasites also transfer from farmed to wild salmon populations. A recent study in British Columbia links the spread of parasitic sea lice from river salmon farms to wild pink salmon in the same river." The European Commission (2002) concluded "The reduction of wild salmonid abundance is also linked to other factors but there is more and more scientific evidence establishing a direct link between the number of lice-infested wild fish and the presence of cages in the same estuary." It is reported that wild salmon on the west coast of Canada are being driven to extinction by sea lice from nearby salmon farms. Antibiotics and pesticides are often used to control the diseases and parasites.

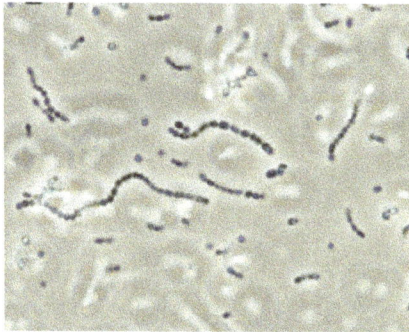

*Streptococcus iniae,* a Gram-positive, sphere-shaped bacteria caused losses in farmed marine and freshwater finfish of US$100 million in 1997.

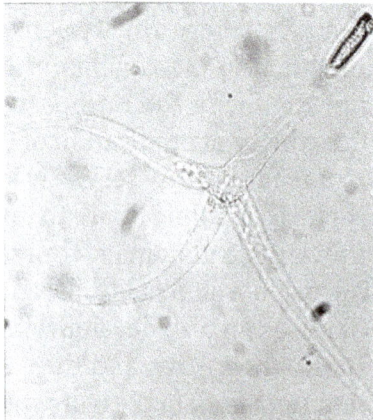

*Myxobolus cerebralis,* a myxosporean parasite, causes *whirling disease* in farmed salmon and trout and also in wild fish populations.

*Henneguya salminicola,* a protozoan parasite commonly found in the flesh of salmonids on the West Coast of Canada. Coho salmon

Columnaris in the gill of a chinook salmon

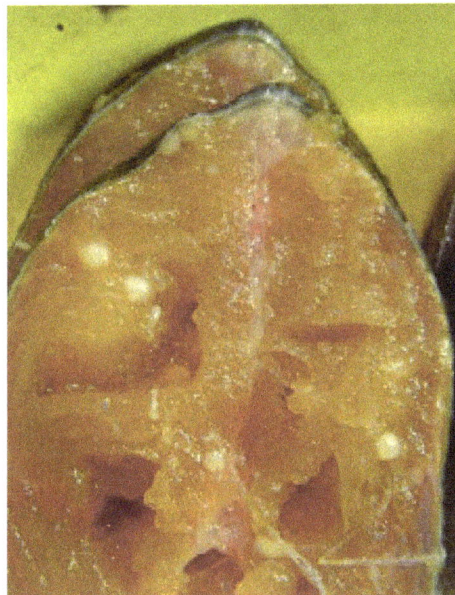

The parasite *Henneguya zschokkei* in salmon beard

- Sockeye salmon with gas bubble disease

## References

- Axelrod HR, Untergasser D (1989). Handbook of fish diseases. Neptune, NJ: T.F.H. Publications. ISBN 0-86622-703-2.

- Exell A, Burgess PH, Bailey MT. A-Z of Tropical Fish Diseases and Health Problems. New York, N.Y: Howell Book House. ISBN 1-58245-049-8.

- Fairfield, T (2000). A commonsense guide to fish health. Woodbury, N.Y: Barron's Educational Series. ISBN 0-7641-1338-0.

- Moyle, PB and Cech, JJ (2004) Fishes, An Introduction to Ichthyology. 5th Ed, Benjamin Cummings. ISBN 978-0-13-100847-2

- Woo PTK (1995) Fish Diseases and Disorders: Volume 1: Protozoan and Metazoan Infections Cabi Series. ISBN 9780851988238.

- Woo PTK (2011) Fish Diseases and Disorders: Volume 2: Non-Infectious Disorders Cabi Series. ISBN 9781845935535.

- Woo PTK (2011) Fish Diseases and Disorders: Volume 3: Viral, Bacterial and Fungal Infections Cabi Series. ISBN 9781845935542.

- Axelrod HR and Untergasser D (1989). Handbook of fish diseases. Neptune, NJ: T.F.H. Publications. ISBN 0-866227032.

- Fairfield, T (2000). A commonsense guide to fish health. Woodbury, N.Y: Barron's Educational Series. ISBN 0-7641-1338-0.

# Permissions

# Index

www.ingramcontent.com/pod-product-compliance
Lightning Source LLC
Chambersburg PA
CBHW082049190326
41458CB00010B/3494